PRACTICAL ELECTRONICS CALCULATIONS AND FORMULAE

ALSO BY THE SAME AUTHOR

BP144 FURTHER PRACTICAL ELECTRONICS CALCULATIONS AND FORMULAE

BP286 A REFERENCE GUIDE TO BASIC ELECTRONICS TERMS

BP287 A REFERENCE GUIDE TO PRACTICAL ELECTRONICS TERMS

PRACTICAL ELECTRONICS CALCULATIONS AND FORMULAE

by
F.A. WILSON
C.G.I.A., C.Eng., F.I.E.E., F.I.E.R.E., F.B.I.M.

BERNARD BABANI (publishing) LTD
THE GRAMPIANS
SHEPHERDS BUSH ROAD
LONDON W6 7NF
ENGLAND

PLEASE NOTE

Although every care has been taken with the production of this book to ensure that any projects, designs, modifications and/or programs etc. contained herewith, operate in a correct and safe manner and also that any components specified are normally available in Great Britain, the Publishers do not accept responsibility in any way for the failure, including fault in design, of any project, design, modification or program to work correctly or to cause damage to any other equipment that it may be connected to or used in conjunction with, or in respect of any other damage or injury that may be so caused, nor do the Publishers accept responsibility in any way for the failure to obtain specified components.

Notice is also given that if equipment that is still under warranty is modified in any way or used or connected with home-built equipment then that warranty may be void.

The cover illustration may not represent a formula contained in this book.

I.S.B.N. 0 900162 70 8

First Published – January 1979
Revised and Reprinted – December 1981
Reprinted – October 1987
Reprinted – January 1991
Reprinted – May 1993

Printed and bound in Great Britain by Cox & Wyman Ltd, Reading.

PREFACE

This booklet has been written, not for the family bookshelves, but for the electronic enthusiast's workshop bench. Its aim is to bridge the gap between complicated technical theory which sometimes seems to have little relevance to practical work and "cut and try" methods which may bring success in design but leave the experimenter unfulfilled.

There is therefore a strong practical bias, tedious and higher mathematics have been avoided where possible and many tables have been included, partly to save calculation and partly because actual figures bring a greater intimacy with the design process.

Yet for those who in their technical studies have found difficulty, or in consort with most other humans have lapses of memory, there is plenty of help and revision.

As a reference book, sections have been written to be as self-contained as possible.

F.A. Wilson, C.G.I.A., C.Eng., F.I.E.E., F.I.E.R.E., F.B.I.M.

CONTENTS

Page

Page

1. UNITS AND CONSTANTS

1.1 UNITS

1.1.1 Basic S.I. Units

These are the metric units in the Système International d'Unités (S.I.)

Quantity	*Name of Unit*	*Symbol*
Length	metre	m
Mass	kilogramme	kg
Time	second	s
Electric Current	ampere	A
Temperature	degree kelvin	°K
Luminous Intensity	candela	cd

1.1.2 Derived S.I. Units

Some derived units more commonly met in electronic engineering

Quantity	*Name of Unit*	*Symbol*
Force	newton	$N = kg\ m/s^2$
Work, Energy, Quantity of Heat	joule	$J = Nm$
Power	watt	$W = J/s$
Electric Charge	coulomb	$C = As$
Electric Potential	volt	$V = W/A$
Electric Capacitance	farad	$F = As/V$
Electric Resistance	ohm	$\Omega = V/A$
Frequency	hertz	$Hz = s^{-1}$
Magnetic Flux	weber	$Wb = Vs$
Magnetic Flux Density	tesla	$T = Wb/m^2$
Inductance	henry	$H = Vs/A$

1.2 ABBREVIATIONS AND SYMBOLS

1.2.1 Multiples and Sub-Multiples of Units

Multiplication Factor	*Prefix*	*Symbol*
10^{12}	tera	T
10^{9}	giga	G
10^{6}	mega	M
10^{3}	kilo	k
10^{2}	hecto	h
10^{1}	deca	da
10^{-1}	deci	d
10^{-2}	centi	c
10^{-3}	milli	m
10^{-6}	micro	μ
10^{-9}	nano	n
10^{-12}	pico	p
10^{-15}	femto	f
10^{-18}	atto	a

1.2.2 Mathematical Symbols

A few, perhaps lesser-known, symbols used in this book are shown below:

$\simeq$	approximately equal to
$\neq$	not equal to
$\equiv$	is identical with
$\propto$	varies as, proportional to
$>$	greater than
$<$	less than
∞	infinity
j	complex operator, $\sqrt{-1}$
$\|Z\|$	modulus of complex number (Z)

1.2.3 Greek Alphabet

Name	*Lower Case (small letter)*	*Capital Letter*
Alpha	α	A
Beta	β	B
Gamma	γ	Γ
Delta	δ	Δ
Epsilon	ϵ	E
Zeta	ζ	Z
Eta	η	H
Theta	θ	Θ
Iota	ι	I
Kappa	κ	K
Lambda	λ	Λ
Mu	μ	M
Nu	ν	N
Xi	ξ	Ξ
Omicron	o	O
Pi	π	Π
Rho	ρ	P
Sigma	σ	Σ
Tau	τ	T
Upsilon	υ	Υ
Phi	ϕ	Φ
Chi	χ	X
Psi	ψ	Ψ
Omega	ω	Ω

1.3 CONSTANTS

This list is not exhaustive, it relates mainly to the subject of this book.

$\pi = 3.14159265$ $\quad \dfrac{1}{\pi} = 0.31830989$ $\quad \pi^2 = 9.8696044$

$\sqrt{\pi} = 1.77245385$

$\log_{10} \pi = 0.4971$ $\quad \log_{10} \dfrac{1}{\pi} = \bar{1}.5029$ $\quad \log_{10} \pi^2 = 0.9943$

$\log_{10} \sqrt{\pi} = 0.2486$

$e = 2.71828183$ $\quad \dfrac{1}{e} = 0.36787944$

$\log_{10} e = 0.4343$

$\log_{10} \frac{1}{e} = \bar{1}.5657$

Radian = $57° \; 17' \; 44.8''$

Radians	*Degrees*
1	57.30
2	114.59
3	171.89
4	229.18
5	286.48
6	343.77
2π	360

Velocity of light/radio waves

= 299774 km/sec $\simeq 3 \times 10^8$ metres/sec
= 186271 miles/sec $\simeq$ 186000 miles/sec

1.4 CONVERSIONS

UK units in terms of SI (metric) units

Length	1 in	25.4 mm
	1 ft	304.8 mm
	1 yd	0.9144 m
	1 mile	1.609344 km
Area	1 in^2	645.16 mm^2
	1 ft^2	0.092903 m^2
	1 yd^2	0.836127 m^2
	1 $mile^2$	2.58999 km^2
Volume	1 in^3	16387.1 mm^3
	1 ft^3	0.0283168 m^3
	1 gallon	4.54609 dm^3 $\simeq$ 4.546 litres
Velocity	1 ft/s	0.3048 m/s
	1 mile/h	0.44704 m/s

Mass	1 lb	0.45359237 kg
Power	1 hp	745.7 W
Energy	1 calorie	4.1868 J
	1 Btu	1.05506 kJ
	1 kWh	3.6×10^6 J

2. DIRECT CURRENT CIRCUITS

2.1 DEFINITIONS

The Unit of Charge is the COULOMB which is defined as the quantity of electricity (Q) passing a given point in a circuit in one second when the current (I) is one ampere.

$$\text{Hence} \quad I\,(\text{amperes}) = \frac{Q\,(\text{coulombs})}{t\,(\text{seconds})}$$

The Unit of Resistance is the OHM which is defined as that resistance in which a current of one ampere flowing for one second generates one joule of heat energy.

Since the heat generated is proportional to the square of the current,

$$R = \frac{\text{Heat generated in joules}}{I^2 t}\ \text{ohms}$$

where I = current in amperes
t = time in seconds
R = resistance in ohms

OHM's LAW relates potential difference (V), current and resistance

$$\frac{V}{I} = R$$

V is in volts
I is in amperes
R is in ohms

$$\text{Hence,} \quad I = \frac{V}{R}, \qquad V = IR$$

The unit of work is the joule and if a charge of electricity of Q coulombs is moved through a p.d. of V volts, then the work done,

$$W\,(\text{joules}) = Q \times V$$

Power is the rate of doing work

$$\text{ie } \frac{W}{t} \text{ where t = time in seconds}$$

$$\therefore \text{ Power, P (watts)} = \frac{QV}{t}$$

and since

$$I = \frac{Q}{t}$$

$$\text{P (watts)} = \text{I (amps)} \times \text{V (volts)}$$

and from Ohm's Law

$$P = \frac{V^2}{R} = I^2R \qquad \text{where R = resistance of circuit in ohms}$$

EXAMPLE:
A 12V car headlamp bulb is rated at 36 watts. What current does it take and what is its working resistance?

$$I = \frac{P}{V} = \frac{36}{12} = 3 \text{ amps}$$

$$R = \frac{V^2}{P} \text{ or } \frac{P}{I^2} \text{ or } \frac{V}{I} = 4 \text{ ohms}$$

2.2 PRIMARY AND SECONDARY CELLS

A primary cell produces electrical energy by chemical action by which one of the electrodes is consumed. Generally, it cannot be re-charged. A secondary cell produces energy by changes in the chemical composition of the electrodes. The changes can be reversed by charging, i.e. passing a current through the cell in the opposite direction to the discharge current.

Representation of a battery on a diagram (Fig.2.1):

All cells or batteries (a battery consists of more than one cell) have "internal resistance" represented by r in the diagram.

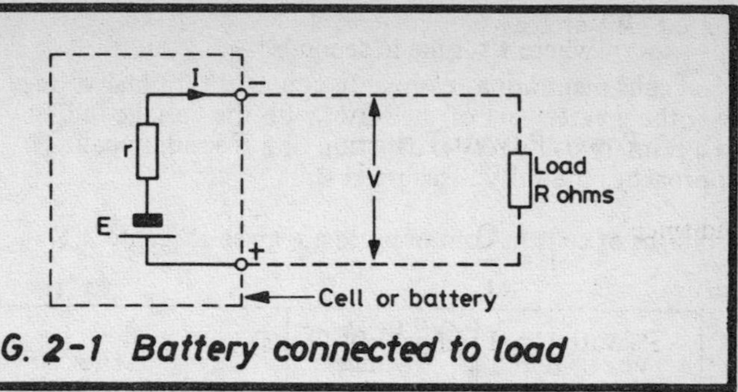

FIG. 2-1 Battery connected to load

E represents the electromotive force (e.m.f.) of the battery which is assumed to be constant. V is the battery voltage, i.e. the p.d. across its terminals.

With no load connected to terminals, $I = O$, $V = E$

With load connected, current flows:

$$I = \frac{E}{R + r} \text{ and battery voltage V falls to } V = E - Ir$$

2.2.1 Connexion of Cells in Series and Parallel

When n similar cells of e.m.f. E and internal resistance r are connected in series aiding (positive of one cell connected to negative of next) the battery e.m.f. = nE and the internal resistance = nr.

When n similar cells as above are connected in parallel (all positive terminals connected together and all negatives), the battery e.m.f. = E and internal resistance = r/n.

Hence, series connexions produce higher battery voltages e.g. 4 1.5V dry cells make a 6V battery, but parallel connexions have the same voltage as a single cell but lower internal resistance. The capacity is, of course, greater.

2.2.2 Cell Voltages

Most cells maintain a reasonably constant terminal voltage over the greater part of their life, with the voltage falling as a primary cell nears exhaustion or a secondary cell approaches the fully discharged state.

Voltages of cells in common use are given in Table 2.1.

	Type of Cell	*Cell Working Voltage (approx.)*	*Use*
Primary	Leclanché – "dry battery"	1.5	Universal – torches, transistor radios, calculators, etc.
Primary	Mercury	1.3	Hearing aids, watches
Secondary	Lead-Acid	2.0	Where Charge/Discharge facility required – vehicles, standby batteries, etc.
Secondary	Alkaline (Nickel-Iron)	1.2	Heavy Duty, traction work.
Secondary	Nickel-Cadmium	1.2	Electronic, photographic applications etc.

Table 2.1 **Cell Voltages and Uses**

2.3 ELECTROMAGNETISM

2.3.1 Useful Rules for the Workshop

(1) Magnet poles:

(i) the North Pole of a magnet is more correctly described as the "North Seeking Pole" because the magnet, when freely suspended, will take up a position with its N pole pointing to the Earth's North Magnetic Pole. The South seeking Pole of the magnet must therefore point to the Earth's South Magnetic Pole.

(ii) Like poles repel, unlike poles attract.

(iii) (i) and (ii) above lead to the proposition that the Earth itself behaves as a bar magnet with its North Magnetic Pole actually a South Pole and vice versa in the terminology that is in use for labelling magnets.

(2) Maxwell's Corkscrew Rule:

If a corkscrew is visualized as being screwed along the wire in the direction of the current (conventional current flow, + to –), then the direction of the magnetic flux around the wire is the same as the direction of rotation of the corkscrew. Diagrammatically, a cross represents the tail end of an arrow and indicates current flowing away from the observer, while a dot represents the point of the arrow and indicates current flowing towards the observer. Thus Maxwell's rule is illustrated by both (a) and (b) in Fig.2.2.

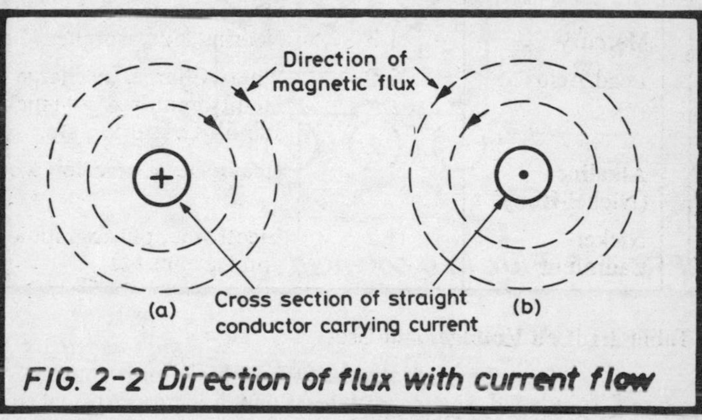

FIG. 2-2 Direction of flux with current flow

(3) Magnetic Polarity of a Solenoid:

Fig.2.3(a) represents a simple solenoid of 3 turns only with the direction of current marked, (b) shows a vertical section of the solenoid with magnetic flux directions added (only one "line of force" is shown for clarity). The direction of the main solenoid flux adds up as shown and within the solenoid is left to right with the magnetic circuit outside the solenoid being completed with flux from right to left.
Since flux is considered to flow from N to S outside of a magnet, the solenoid must have magnetic polarity as marked.

Equally the corkscrew rule can be used by imagining the

corkscrew being screwed along the line of progression of the current, the direction of the corkscrew then indicating the direction of flux within the solenoid.

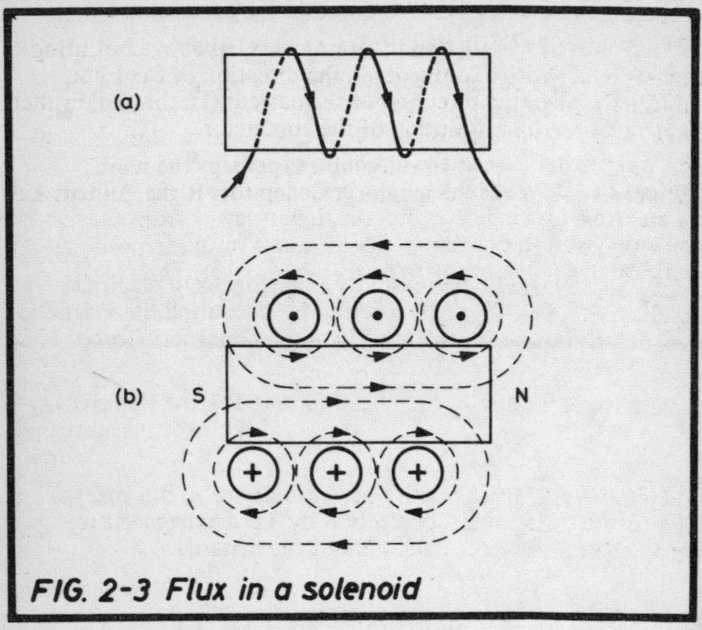

FIG. 2-3 Flux in a solenoid

(4) Faraday's Law:

When the magnetic flux through a circuit changes, an e.m.f. is induced, the magnitude of which is proportional to the rate of change of flux.

(5) Lenz's Law follows:

The e.m.f. induced in any circuit is always in such a direction that its effect tends to oppose the motion or change producing it.

(6) Fleming's Rules:

(a) Right Hand – for Generators ie. a conductor being moved through, and at right angles to, a magnetic field. The thumb and middle finger of the right hand are extended so that, with the forefinger, the three are all at right angles to each other (as the edges of the sides, meeting at a corner of a box). If the hand is then turned so that the thuMb points in the direction of the Motion of the conductor and the ForeFinger in the direction of the Flux then thE Middle Finger will point in the direction of the EMF induced in the conductor.

(b) Left Hand – for Motors. As above but using the left hand, with ForeFinger in the direction of the Flux, mIddle fInger in the direction of the current (I), the thuMb then gives the direction of Motion of the conductor.

It is fatal to get these two rules mixed, perhaps the word GRuMbLe might aid the memory, Generators Right, Motors Left.

(7) Permeability:

Magnetic permeability is analogous to electrical conductivity, and is usually assessed by measuring the magnetic flux density set up in a material by a given magnetic force.

ie, relative permeability $(\mu_r) = \dfrac{B}{H}$ where B is the flux density, H is the magnetizing force

and if a material has a relative permeability of x, this means that the flux density produced in it by a given magnetizing force is x times greater than without the material.

Note

As mentioned earlier the rules in (2), (3) and (6) are for conventional current flow. If based on the modern concept of electron flow then the directions of magnetic flux in Figs. 2.2 and 2.3 are reversed and Fleming's Rules as originally conceived do not apply although the same technique is applicable using the right hand for motors and the left for generators.

3. PASSIVE COMPONENTS

3.1 RESISTANCE

3.1.1 Resistivity

Variation of Resistance with Dimension

The resistance of a conductor of uniform gauge is directly proportional to its length and inversely proportional to its cross-sectional area.

If length is denoted by l, diameter by d and cross-sectional area by a,

$$\text{Resistance (R)} \propto \frac{l}{a}, \text{ also } R \propto \frac{l}{d^2}$$

Variation of Resistance with Material

Each material has its own "volume resistivity" or "specific resistance" ("volume resistivity" is the preferred term but for brevity, the term "resistivity" is used here), denoted by ρ and accordingly

$$R = \rho\frac{l}{a}$$

ρ is defined as the resistance between opposite faces of a metre cube of the material as shown in Fig.3.1.

whence l = 1 metre
a = 1 sq metre
ie R = ρ and in this case
ρ is expressed in "ohm-metres"

The ohm-metre is not a very convenient unit for conductors because of its very low value and frequently the microhm-metre or microhm-centimetre is used. Note that in the formula, whichever unit is used for ρ, then both l and a must be in the same system of units.

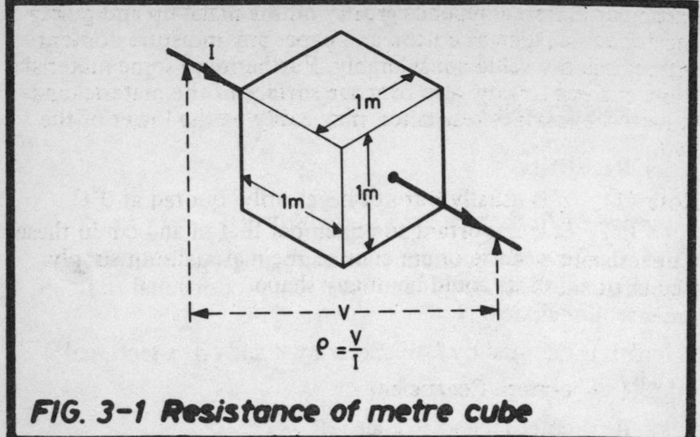

FIG. 3-1 Resistance of metre cube

Units in use for ρ with their equivalents are:

1 ohm-metre	=	10^6 microhm-metres
1 ohm-metre	=	10^8 microhm-centimetres
1 microhm-metre	=	39.4 microhm-inches
1 microhm-metre	=	100 microhm-centimetres
1 Megohm-metre	=	10^6 ohm-metres

Megohm-metres are used for insulating materials and the order of value of some commonly used materials for electronic work in Megohm-metres are

Air (dry) – Infinite
Ceramics and plastics – 10^{10}
Paraffin Wax – 10^9
Mica – $10^8 - 10^9$
Rubber – 10^8
Shellac – $10^7 - 10^8$
Porcelain – $10^4 - 10^7$
Cotton, paper, silk – $10^4 - 10^6$
Glass – 10^5

It must be emphasized that these figures are very approximate and show the order of resistivities only. The actual figure for a particular material depends greatly on the make-up and purity and for some, such as cotton and paper any moisture content will reduce the value considerably. Furthermore some materials allow current to flow also over the surface of the material and in fact, the 'surface resistance' may easily be the lower of the two.

Note (1) ρ is usually but not necessarily, quoted at 0°C
(2) it is important to remember that m and cm in these units refer to a metre or cm cube of the material, not simply a cu.m or cc, these could be of any shape.

3.1.2 Temperature Coefficient

Variation of Resistance with Temperature

This has been determined for most electrical materials and when expressed as the fractional change per degree Centigrade, it is known as the Temperature Coefficient of Resistance, symbol a. If R_t is the resistance of a conductor at t°C, and R_o the resistance at 0°C,

$$\text{Then } R_t = R_o (1 + at + \beta t^2)$$

where a is the coefficient depending on the particular material. β is a second coefficient but it can be shown that it can be neglected for most work, especially when the temperature does not greatly exceed 100°C. In fact β for copper is +0.00000112 and neglecting this term results in only 0.04% error at 20°C rising to about 0.8% at 100°C.

$$\text{Hence } R_t \simeq R_o (1 + at)$$

and if resistance = R_1 at t_1°C and R_2 at t_2°C

$$\text{then } \frac{R_1}{R_2} = \frac{1 + at_1}{1 + at_2}$$

a is positive when R increases with t, negative when R decreases with increase of t.

3.1.3 Calculation of Resistance

Values of ρ and a for some commonly used materials are given in Table 3.1.

Material	ρ ($\mu\Omega$/cm) at 0°C	a	ρ'
Silver	1.48	+0.0037	0.0202
Copper	1.60	+0.0039	0.0220
Aluminium	2.62	+0.0038	0.0359
Carbon	3535	–0.0005	44.55
Constantan	49.0	+0.000014	0.6239
Eureka	49.0	+0.000014	0.6239
Nickel	9.12	+0.0059	0.1298
Nickel Silver 1	30.8	+0.00027	0.3942
Nickel Silver 4	20.8	+0.00047	0.2673
Nicrome 5%	89.1	+0.00105	1.1580
Nicrome 15%	109.6	+0.0002	1.4008
Manganin	44.0	+0.000014	0.5603
Platinoid	41.75	+0.0003	0.5347

For ρ in $\mu\Omega$/inch, divide values in table by 2.54

Table 3.1 **Resistivity and Temperature Coefficient**

EXAMPLE:
What is the resistance of 10 metres of copper wire of diameter 0.2mm (approx. SWG 36) at 20°C?

Cross-sectional area $= \pi r^2 = \pi \times 10^{-2}$ sq mm $= \pi \times 10^{-4}$ sq cm

$$l = 10 \times 10^2 = 10^3 \text{ cm}$$

From table $\rho = 1.60\ \mu\Omega/\text{cm}$

$$\text{Then } R_o\ (0°\text{C}) = \frac{1.60 \times 10^3}{\pi \times 10^{-4}}\ \mu\Omega = \frac{1.60}{\pi} \times 10\Omega = 5.092\Omega$$

$$\text{and } R_{20}\ (20°\text{C}) = 5.092\,(1 + 0.0039 \times 20) = 5.50\Omega$$

3.1.4 Resistance of Wires

Because consideration of a practical range of wire sizes in each of the materials results in a multitude of tables, the following

Diameter (d) mm	d^2
0.1	0.01
0.11	0.0121
0.12	0.0144
0.14	0.0196
0.16	0.0256
0.18	0.0324
0.20	0.04
0.22	0.0484
0.24	0.0576
0.26	0.0676
0.28	0.0784
0.30	0.09
0.35	0.123
0.40	0.160
0.45	0.203
0.50	0.250
0.55	0.303
0.6	0.36
0.7	0.49
0.8	0.64
0.9	0.81
1.0	1.0
1.1	1.21
1.2	1.44
1.4	1.96
1.6	2.56
1.8	3.24
2.0	4.0

S.W.G.	d^2
15	3.345
16	2.644
17	2.022
18	1.486
19	1.032
20	0.836
21	0.661
22	0.506
23	0.372
24	0.312
25	0.258
26	0.209
27	0.173
28	0.141
29	0.119
30	0.0992
31	0.0868
32	0.0752
33	0.0645
34	0.0546
35	0.0455
36	0.0373
37	0.0298
38	0.0232
39	0.0175
40	0.0149
41	0.0125
42	0.0103

Table 3.2 **Values of d^2 for Wire Resistance Calculations**

method is given, by which, through a small calculation the resistance of wires for any material quoted in the table, can be quickly obtained. For materials not quoted it is also a straightforward calculation provided that ρ and α are known. Generally, wire resistance values are required at "room temperature" which here is taken to be 20°C.

The last column of Table 3.1 gives values for ρ', these figures have been calculated so that when simply divided by d (the wire diameter in mm) squared, the resistance of the wire per metre at 20°C is given. To simply further, Table 3.2 gives values of d^2 for a practical range of wire sizes in both metric and Standard Wire Gauge form. The results of this method are accurate to within about 0.25%.

EXAMPLE:
What is the resistance per metre of (1) SWG 40 copper wire, (2) 0.5 mm diameter Eureka wire?

(1) From Table 3.1 ρ' for copper = 0.022
From Table 3.2 d^2 for SWG 40 = 0.0149

$$\therefore \text{Resistance/metre} = \frac{\rho'}{d^2} = \frac{0.022}{0.0149} = 1.4765 \text{ ohms}$$

(2) ρ' for Eureka = 0.6239
d^2 for 0.5 mm = 0.25

$$\therefore \text{Resistance/metre} = \frac{0.6239}{0.25} = 2.4956 \text{ ohms}$$

3.2 RESISTORS

3.2.1 Coding and Preferred Values

The Colour Code for Resistors is given in Table 3.3 and Fig.3.2 shows the method of reading on two commonly used types.

Colour	*Used as Significant Figure*	*Used as Decimal Multiplier*
Silver		0.01
Gold		0.1
Black	0	1
Brown	1	10
Red	2	10^2
Orange	3	10^3
Yellow	4	10^4
Green	5	10^5
Blue	6	10^6
Violet	7	10^7
Grey	8	10^8
White	9	10^9

Tolerances: Silver, 10%, Gold, 5%.

Table 3.3 **Colour Code for Resistors and Capacitors**

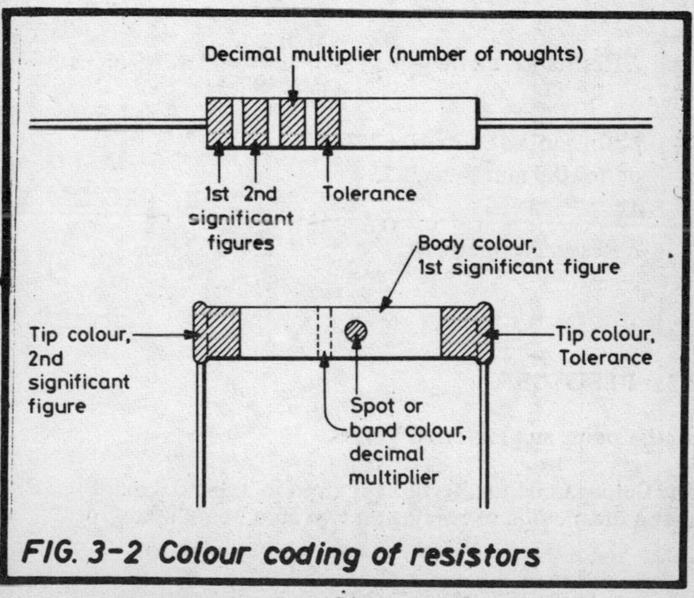

FIG. 3-2 Colour coding of resistors

TOLERANCE	BASIC SERIES OF VALUES																								SERIES
20%	10				15				22				33				47				68				E6
10%	10		12		15		18		22		27		33		39		47		56		68		82		E12
5%	10	11	12	13	15	16	18	20	22	24	27	30	33	36	39	43	47	51	56	62	68	75	82	91	E24

Table 3.4 Preferred Values for Resistors and Capacitors

The Preferred Value System has been designed for minimum overlapping between one value and the next due to tolerance spreads and the three series in use are given in Table 3.4.

(Also see Appendix for other marking codes.)

3.2.2 Series Combinations

Consider Fig.3.3 in which the same current I is flowing through two resistances R_1 and R_2 in series.

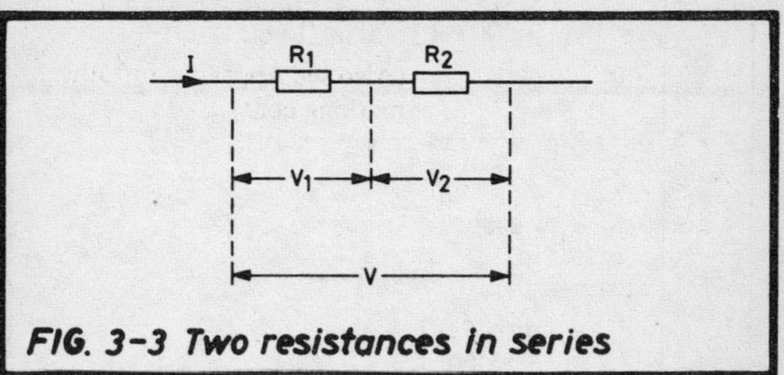

FIG. 3-3 Two resistances in series

Let V_1 and V_2 be the voltages developed as shown and V the total voltage

$$\text{ie} \quad V = V_1 + V_2$$

Now from Ohm's Law

$$IR_1 = V_1 \quad \text{and} \quad IR_2 = V_2$$

$$\therefore V_1 + V_2 = V = I(R_1 + R_2)$$

$$\therefore \frac{V}{I} = (R_1 + R_2) = R, \text{ the total resistance of the combination}$$

Thus the total resistance of two resistances in series is equal to their sum and this rule can be shown to apply similarly for any number of resistances in series, ie

$$R = R_1 + R_2 + R_3 + \ldots$$

EXAMPLE:
A biassing circuit consists of two resistors connected across a 9 V supply as shown in Fig.3.4. What is the voltage at point A relative to the negative line?

Total Resistance $R = R_1 + R_2 = 9000$ ohms

By Ohm's Law $I = \dfrac{9}{9000} = .001$ amps

$\therefore V_2 = IR_2 = .001 \times 1000 = 1$ volt

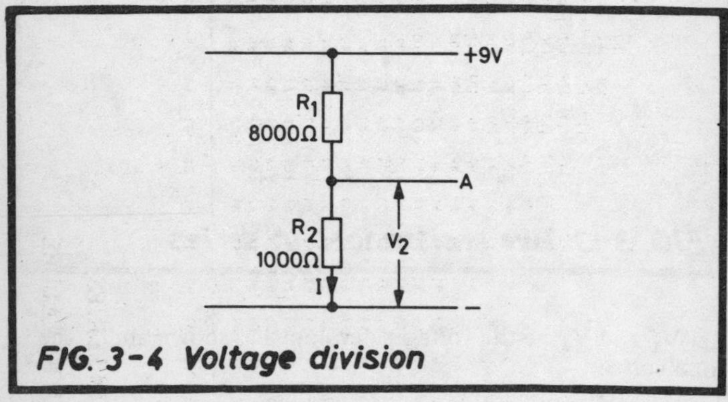

FIG. 3-4 Voltage division

Two Preferred Value Resistors in Series

Although it is comparatively easy to find two preferred values which, when connected in series make up a particular required value, Table 3.5 gives this in a convenient form as a reminder of all choices which can be made. It covers a complete decade of values for R and it is of interest that exact values for all numbers between 20 and 100 (with the single exception of 96) can be found from the 5% tolerance range. Higher or lower ranges are obtained by multiplying or dividing by multiples of 10. For some values several choices exist, eg 63 ohms can be obtained from (51 + 12), (47 + 16), (43 + 20), (39 + 24), (36 + 27) or (33 + 30).

It is also worth remembering that if both R_1 and R_2 have a certain tolerance, then the value of the series combination, R, will have the same tolerance. The general formula where t, t_1 and t_2 are the appropriate tolerances in percentage terms is:

$$R_t = R_1 t_1 + R_2 t_2$$

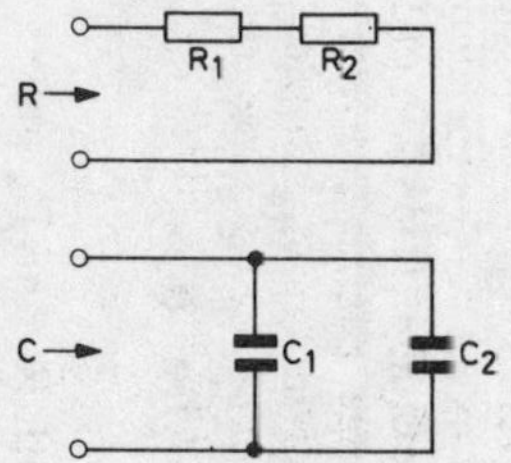

R₁ or R₂ / C₁ or C₂

R_1 or R_2 / C_1 or C_2	10	11	12	13	15	16	18	20	22	24	27	30	33	36	39	43	47	51	56	62	68	75	82	91	100	R_1 or R_2 / C_1 or C_2
100																									200	100
91																								182	191	91
82																							164	173	182	82
75																						150	157	166	175	75
68																					136	143	152	159	168	68
62																				124	130	137	146	153	162	62
56																			112	118	124	131	140	147	156	56
51																		102	107	113	119	126	135	142	151	51
47																	94	98	103	109	115	122	129	138	147	47
43																86	89	94	99	105	111	118	125	134	143	43
39															78	82	85	90	95	101	107	114	121	130	139	39
36														72	75	79	83	87	92	98	104	111	118	127	136	36
33													66	69	72	76	80	84	89	95	101	108	115	124	133	33
30												60	63	66	69	73	77	81	86	92	98	105	112	121	130	30
27											54	57	60	63	66	70	74	78	83	89	95	102	109	118	127	27
24										48	51	54	57	60	63	67	71	75	80	86	92	99	106	115	124	24
22									44	46	49	52	55	58	61	65	69	73	78	84	90	97	104	113	122	22
20								40	42	44	47	50	53	56	59	63	67	71	76	82	88	95	102	111	120	20
18							36	38	40	42	45	48	51	54	57	61	65	69	74	80	86	93	100	109	118	18
16						32	34	36	38	40	43	46	49	52	55	59	63	67	72	78	84	91	98	107	116	16
15					30	31	33	35	37	39	42	45	48	51	54	58	62	66	71	77	83	90	97	106	115	15
13				26	28	29	31	33	35	37	40	43	46	49	52	56	60	64	69	75	81	88	95	104	113	13
12			24	25	27	28	30	32	34	36	39	42	45	48	51	55	59	63	68	74	80	87	94	103	112	12
11		22	23	24	26	27	29	31	33	35	38	41	44	47	50	54	58	62	67	73	79	86	93	102	111	11
10	20	21	22	23	25	26	28	30	32	34	37	40	43	46	49	53	57	61	66	72	78	85	92	101	110	10

R_2 or R_1 / C_2 or C_1

TABLE 3.5 VALUES OF { **R FOR R_1 AND R_2 IN SERIES** / **C FOR C_1 AND C_2 IN PARALLEL** }

Hence, tolerance of the combination:

$$t = \frac{R_1 t_1 + R_2 t_2}{R}$$

for example, if the 51 and 12 used to make 63 as above are 5% and 10% respectively, then

$$63t = (51 \times 5) + (12 \times 10)$$

$$\therefore \; t = 5.95\%$$

3.2.3 Parallel Combinations

Fig.3.5 shows two resistances R_1 and R_2 connected in parallel with currents I_1 and I_2 flowing through them respectively.

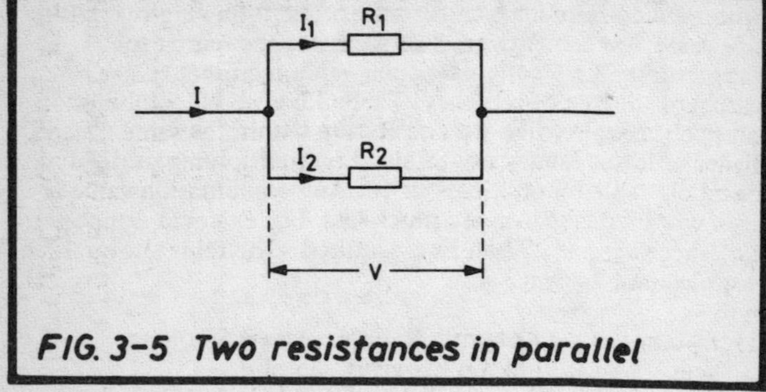

FIG. 3-5 Two resistances in parallel

Let the combined resistance be R ohms. This is the converse of the series case in that the voltage V is the same across both resistances but the current I divides between them.

Then $I = I_1 + I_2$

and by Ohm's Law $\quad I_1 = \dfrac{V}{R_1} \quad$ and $\quad I_2 = \dfrac{V}{R_2}$

$$\therefore I = V\left(\frac{1}{R_1} + \frac{1}{R_2}\right)$$

$$\text{But} \quad \frac{I}{V} = \frac{1}{R} \qquad \therefore \frac{1}{R} = \frac{1}{R_1} + \frac{1}{R_2}$$

It can similarly be shown that

$$\frac{1}{R} = \frac{1}{R_1} + \frac{1}{R_2} + \frac{1}{R_3} + \text{------}$$

For two resistors only in parallel

$$R = \frac{R_1 R_2}{R_1 + R_2}$$

Two Preferred Value Resistors in Parallel

Calculation of the resultant resistance in the parallel case is more time consuming than for series. The following two tables have therefore been designed to facilitate the search for resistors to make up any particular value required, to a reasonable degree of accuracy. Table 3.6 gives all combination values obtainable when R_1 and R_2 are within the same decade. Higher or lower ranges are obtained by multiplying or dividing the whole table by multiples of 10. Any combination value is given within ±2.5%, usually much less. For example, suppose a resistance value of 950 ohms is required. The table shows three possible combinations:

(1) 13 ohms in parallel with 36 ohms gives 9.55 ohms,
hence 1.3 kΩ with 3.6 kΩ gives 955 ohms
(2) Similarly 1.8 kΩ in parallel with 2.0 kΩ gives 947 ohms
(3) and 1.1 kΩ in parallel with 6.8 kΩ gives 947 ohms

and each of these produces the required value within approximately 0.5%.

To increase the choice of resistors which can be selected (note that the values of R_1 and R_2 in Table 3.6 are within the same decade, irrespective of the fact that the required R may be in ohms, kilohms or megohms), the second table, Table 3.7, is given for R_1 and R_2 in adjacent decades, that is, if the smaller of the two lies in the range x to $10x$ ohms, (where

R_1 or R_2 / C_1 or C_2	10	11	12	13	15	16	18	20	22	24	27	30	33	36	39	43	47	51	56	62	68	75	82	91	100	R_1 or R_2 / C_1 or C_2
100																									50.00	100
91																								45.50	47.64	91
82																							41.00	43.13	45.05	82
75																						37.50	39.17	41.11	42.86	75
68																					34.00	35.66	37.17	38.92	40.48	68
62																				31.00	32.43	33.94	35.31	36.88	38.27	62
56																			28.00	29.42	30.71	32.06	33.28	34.67	35.90	56
51																		25.50	26.69	27.98	29.14	30.36	31.44	32.68	33.77	51
47																	23.50	24.46	25.55	26.73	27.79	28.89	29.88	30.99	31.97	47
43																21.50	22.46	23.33	24.32	25.39	26.34	27.33	28.21	29.20	30.07	43
39															19.50	20.45	21.31	22.10	22.99	23.94	24.79	25.66	26.43	27.30	28.06	39
36														18.00	18.72	19.59	20.39	21.10	21.91	22.78	23.54	24.32	25.02	25.80	26.47	36
33													16.50	17.22	17.88	18.67	19.39	20.04	20.76	21.54	22.22	22.92	23.53	24.22	24.81	33
30												15.00	15.71	16.36	16.96	17.67	18.31	18.89	19.53	20.22	20.82	21.43	21.96	22.56	23.08	30
27											13.50	14.21	14.85	15.43	15.95	16.59	17.15	17.65	18.22	18.81	19.33	19.85	20.31	20.82	21.26	27
24										12.00	12.71	13.33	13.89	14.40	14.86	15.40	15.89	16.32	16.80	17.30	17.74	18.18	18.57	18.99	19.35	24
22									11.00	11.48	12.12	12.69	13.20	13.66	14.07	14.55	14.99	15.37	15.79	16.24	16.62	17.01	17.35	17.72	18.03	22
20								10.00	10.48	10.91	11.49	12.00	12.45	12.86	13.22	13.65	14.03	14.37	14.74	15.12	15.45	15.79	16.08	16.40	16.67	20
18							9.00	9.47	9.90	10.29	10.80	11.25	11.65	12.00	12.32	12.69	13.02	13.30	13.62	13.95	14.23	14.52	14.76	15.03	15.25	18
16						8.00	8.47	8.89	9.26	9.60	10.05	10.43	10.78	11.08	11.35	11.66	11.94	12.18	12.44	12.72	12.95	13.19	13.39	13.61	13.79	16
15					7.50	7.74	8.18	8.57	8.92	9.23	9.64	10.00	10.31	10.59	10.83	11.12	11.37	11.59	11.83	12.08	12.29	12.50	12.68	12.88	13.04	15
13				6.50	6.96	7.17	7.55	7.88	8.17	8.43	8.78	9.07	9.33	9.55	9.75	9.98	10.18	10.36	10.55	10.75	10.91	11.08	11.22	11.38	11.50	13
12			6.00	6.24	6.67	6.86	7.20	7.50	7.76	8.00	8.31	8.57	8.80	9.00	9.18	9.38	9.56	9.71	9.88	10.05	10.20	10.34	10.47	10.60	10.71	12
11		5.50	5.74	5.96	6.35	6.52	6.83	7.10	7.33	7.54	7.82	8.05	8.25	8.43	8.58	8.76	8.91	9.05	9.19	9.34	9.47	9.59	9.70	9.81	9.91	11
10	5.00	5.24	5.45	5.65	6.00	6.15	6.43	6.67	6.88	7.06	7.30	7.50	7.67	7.83	7.96	8.11	8.25	8.36	8.48	8.61	8.72	8.82	8.91	9.01	9.09	10
	10	11	12	13	15	16	18	20	22	24	27	30	33	36	39	43	47	51	56	62	68	75	82	91	100	

R_2 or R_1 / C_2 or C_1

TABLE 3.6 VALUES OF { R FOR R_1 AND R_2 IN PARALLEL / C FOR C_1 AND C_2 IN SERIES }

R_1 or R_2 / C_1 or C_2	100	110	120	130	150	160	180	200	220	240	270	300	330	360	390	430	470	510	560	620	680	750	820	910	1000	R_1 or R_2 / C_1 or C_2
100	50.00	52.38	54.55	56.52	60.00	61.54	64.29	66.67	68.75	70.59	72.97	75.00	76.74	78.26	79.59	81.13	82.46	83.61	84.85	86.11	87.18	88.24	89.13	90.10	90.91	100
91	47.64	49.80	51.75	53.53	56.64	53.01	60.44	62.54	64.37	65.98	68.06	69.82	71.33	72.64	73.78	75.11	76.24	77.22	78.28	79.35	80.26	81.15	81.91	82.73	83.41	91
82	45.05	46.98	48.71	50.28	53.02	54.21	56.34	58.16	59.74	61.12	62.90	64.40	65.68	66.79	67.75	68.87	69.82	70.64	71.53	72.42	73.18	73.92	74.55	75.22	75.79	82
75	42.86	44.59	46.15	47.56	50.00	51.06	52.94	54.55	55.93	57.14	58.70	60.00	61.11	62.07	62.90	63.86	64.68	65.38	66.14	66.91	67.55	68.18	68.72	69.29	69.77	75
68	40.48	42.02	43.40	44.65	46.79	47.72	49.35	50.75	51.94	52.99	54.32	55.43	56.38	57.20	57.90	58.71	59.41	60.00	60.64	61.28	61.82	62.35	62.79	63.27	63.67	68
62	38.27	39.65	40.88	41.98	43.87	44.68	46.12	47.33	48.37	49.27	50.42	51.38	52.19	52.89	53.50	54.19	54.77	55.28	55.82	56.36	56.82	57.27	57.64	58.05	58.38	62
56	35.90	37.11	38.18	39.14	40.78	41.48	42.71	43.75	44.64	45.41	46.38	47.19	47.88	48.46	48.97	49.55	50.04	50.46	50.91	51.36	51.74	52.11	52.42	52.75	53.03	56
51	33.77	34.84	35.79	36.63	38.06	38.67	39.74	40.64	41.40	42.06	42.90	43.59	44.17	44.67	45.10	45.59	46.01	46.36	46.74	47.12	47.44	47.75	48.01	48.29	48.53	51
47	31.97	32.93	33.77	34.52	35.79	36.33	37.27	38.06	38.73	39.30	40.03	40.63	41.14	41.57	41.95	42.37	42.73	43.03	43.36	43.69	43.96	44.23	44.45	44.69	44.89	47
43	30.07	30.92	31.66	32.31	33.42	33.89	34.71	35.39	35.97	36.47	37.09	37.61	38.04	38.41	38.73	39.09	39.40	39.66	39.93	40.21	40.44	40.67	40.86	41.06	41.23	43
39	28.06	28.79	29.43	30.00	30.95	31.36	32.05	32.64	33.13	33.55	34.08	34.51	34.88	35.19	35.45	35.76	36.01	36.23	36.46	36.69	36.88	37.07	37.23	37.40	37.54	39
36	26.47	27.12	27.69	28.19	29.03	29.39	30.00	30.51	30.94	31.30	31.76	32.14	32.46	32.73	32.96	33.22	33.44	33.63	33.83	34.02	34.19	34.35	34.49	34.63	34.75	36
33	24.81	25.38	25.88	26.32	27.05	27.36	27.89	28.33	28.70	29.01	29.41	29.73	30.00	30.23	30.43	30.65	30.83	30.99	31.16	31.33	31.47	31.61	31.72	31.85	31.95	33
30	23.08	23.57	24.00	24.38	25.00	25.26	25.71	26.09	26.40	26.67	27.00	27.27	27.50	27.69	27.86	28.04	28.20	28.33	28.47	28.62	28.73	28.85	28.94	29.04	29.13	30
27	21.26	21.68	22.04	22.36	22.88	23.10	23.48	23.79	24.05	24.27	24.55	24.77	24.96	25.12	25.25	25.40	25.53	25.64	25.76	25.87	25.97	26.06	26.14	26.22	26.29	27
24	19.35	19.70	20.00	20.26	20.69	20.87	21.18	21.43	21.64	21.82	22.04	22.22	22.37	22.50	22.61	22.73	22.83	22.92	23.01	23.11	23.18	23.26	23.32	23.38	23.44	24
22	18.03	18.33	18.59	18.82	19.19	19.34	19.60	19.82	20.00	20.15	20.34	20.50	20.63	20.73	20.83	20.93	21.02	21.09	21.17	21.25	21.31	21.37	21.43	21.48	21.53	22
20	16.67	16.92	17.14	17.33	17.65	17.78	18.00	18.18	18.33	18.46	18.62	18.75	18.86	18.95	19.02	19.11	19.18	19.25	19.31	19.38	19.43	19.48	19.52	19.57	19.61	20
18	15.25	15.47	15.65	15.81	16.07	16.18	16.36	16.51	16.64	16.74	16.88	16.98	17.07	17.14	17.21	17.28	17.34	17.39	17.44	17.49	17.54	17.58	17.61	17.65	17.68	18
16	13.79	13.97	14.12	14.25	14.46	14.55	14.69	14.81	14.92	15.00	15.10	15.19	15.26	15.32	15.37	15.43	15.47	15.51	15.56	15.60	15.63	15.67	15.69	15.72	15.75	16
15	13.04	13.20	13.33	13.45	13.64	13.71	13.85	13.95	14.04	14.12	14.21	14.29	14.35	14.40	14.44	14.49	14.54	14.57	14.61	14.65	14.68	14.71	14.73	14.76	14.78	15
13	11.50	11.63	11.73	11.82	11.96	12.02	12.12	12.21	12.27	12.33	12.40	12.46	12.51	12.55	12.58	12.62	12.65	12.68	12.71	12.73	12.76	12.78	12.80	12.82	12.83	13
12	10.71	10.82	10.91	10.99	11.11	11.16	11.25	11.32	11.38	11.43	11.49	11.54	11.58	11.61	11.64	11.67	11.70	11.72	11.75	11.77	11.79	11.81	11.83	11.84	11.86	12
11	9.91	10.00	10.08	10.14	10.25	10.29	10.37	10.43	10.48	10.52	10.57	10.61	10.65	10.67	10.70	10.73	10.75	10.77	10.79	10.81	10.82	10.84	10.85	10.87	10.88	11
10	9.09	9.17	9.23	9.29	9.38	9.41	9.47	9.52	9.57	9.60	9.64	9.68	9.71	9.73	9.75	9.77	9.79	9.81	9.82	9.84	9.86	9.87	9.88	9.89	9.90	10

R_2 or R_1 / C_2 or C_1

TABLE 3.7 VALUES OF {R FOR R_1 AND R_2 IN PARALLEL; C FOR C_1 AND C_2 IN SERIES}

$x = 1, 10, 100$ etc.) then the other is within the range $10x$ to $100x$ ohms. This increases the choices possible for the individual resistors as shown below. Higher or lower ranges are obtained by multiplying or dividing by multiples of 10 as above. Again, for the 950 ohm resistance required:

10 ohms in parallel with 180 ohms gives 9.47 ohms,
hence 1 kΩ in parallel with 18 kΩ gives 947

and similarly 1 kΩ in parallel with 20 kΩ gives 952

so that the two tables together immediately suggest five separate pairs of preferred value resistors which in parallel make up this particular required value to an accuracy within 0.5%. Equally, several choices of combination will be found for any other value of R.

The tables do not cover every combination possible; combination values may also be obtained by the practice of 'padding,' that is, by taking a resistor of value close to that which is required and adjusting this by adding a low value resistor in series or a high value resistor in parallel e.g. 950 ohms is obtained within a few ohms by taking the nearest lower preferred value of 910 and adding a 39 in series, or the nearest higher preferred value of 1 kΩ with an 18 kΩ in parallel (but note that the latter combination has already been suggested in Table 3.7).

As in the series case, if both R_1 and R_2 have a certain tolerance, then the value of the parallel combination, R, will have the same tolerance. The general formula where t, t_1 and t_2 are the appropriate tolerances in percentage terms is:

$$\frac{t}{R} = \frac{t_1}{R_1} + \frac{t_2}{R_2}$$

Hence, tolerance of combination:

$$t = R\left(\frac{t_1}{R_1} + \frac{t_2}{R_2}\right)$$

3.2.4 Division of Voltage and Current in 2-Resistor Combinations

Series:
(Fig. 3.3)

$$\text{Voltage across } R_1 = V \frac{R_1}{R_1 + R_2}$$

$$\text{Voltage across } R_2 = V \frac{R_2}{R_1 + R_2}$$

Parallel:
(Fig. 3.5)

$$\text{Current through } R_1 = I \frac{R_2}{R_1 + R_2}$$

$$\text{Current through } R_2 = I \frac{R_1}{R_1 + R_2}$$

3.2.5 Calculation of Power Dissipation

To avoid overheating, power ratings of resistors for transistor and other circuitry should not be exceeded. Table 3.8 gives the maximum voltage across and maximum current through, a range of preferred value resistors for four separate power ratings. All other resistor values can be covered on the basis that multiplication of the resistance value by multiples of 100 requires multiplication of the appropriate value of voltage or division of the appropriate value of current by an equal multiple, but of 10.

For example:
(1) What is the maximum working current through a 270 kΩ, ¼ watt resistor? From the table, the maximum current through a 27 Ω resistor so as not to exceed a power dissipation of ¼ watt is 96 mA, therefore the maximum current through a 270 kΩ resistor is

$$96 \div 10 \div 10 = 0.96 \text{ mA.}$$

(2) What is the maximum voltage in the above case? From the table, maximum voltage for 27 ohms is 2.60

∴ maximum voltage for 270 kΩ is

$$2.60 \times 10 \times 10 = 260\text{v.}$$

Resistance (ohms)	1/8 watt		1/4 watt		1/2 watt		1 watt	
	V volts	I mA	V volts	I mA	V volts	I mA	V volts	I mA
10	1.12	112	1.58	158	2.24	224	3.16	316
11	1.17	107	1.66	151	2.35	213	3.32	302
12	1.22	102	1.73	144	2.45	204	3.46	289
13	1.27	98	1.80	139	2.55	196	3.61	277
15	1.37	91	1.94	129	2.74	183	3.87	258
16	1.41	88	2.00	125	2.83	177	4.00	250
18	1.50	83	2.12	118	3.00	167	4.24	236
20	1.58	79	2.24	112	3.16	158	4.47	224
22	1.66	75	2.35	107	3.32	151	4.69	213
24	1.73	72	2.45	102	3.46	144	4.90	204
27	1.84	68	2.60	96	3.67	136	5.20	192
30	1.94	65	2.74	91	3.87	129	5.48	183
33	2.03	62	2.87	87	4.06	123	5.75	174
36	2.12	59	3.00	83	4.24	118	6.00	167
39	2.21	57	3.12	80	4.42	113	6.25	160
43	2.32	54	3.28	76	4.64	108	6.56	153
47	2.42	52	3.43	73	4.85	103	6.86	146
51	2.52	50	3.57	70	5.05	99	7.14	140
56	2.65	47	3.74	67	5.29	94	7.48	134
62	2.79	45	3.94	64	5.57	90	7.87	127
68	2.92	43	4.12	61	5.83	86	8.25	121
75	3.06	41	4.33	58	6.12	82	8.66	115
82	3.20	39	4.53	55	6.40	78	9.06	110
91	3.37	37	4.77	52	6.75	74	9.54	105
100	3.54	35	5.00	50	7.07	71	10.00	100
110	3.71	34	5.24	48	7.42	67	10.49	95
120	3.87	32	5.48	46	7.75	65	10.95	91
130	4.03	31	5.70	44	8.06	62	11.40	88
150	4.33	29	6.12	41	8.66	58	12.25	82
160	4.47	28	6.32	40	8.94	56	12.65	79
180	4.74	26	6.71	37	9.49	53	13.42	75
200	5.00	25	7.07	35	10.00	50	14.14	71
220	5.24	24	7.42	34	10.49	48	14.83	67
240	5.48	23	7.75	32	10.95	46	15.49	65
270	5.81	22	8.22	30	11.62	43	16.43	61
300	6.12	20	8.66	29	12.25	41	17.32	58
330	6.42	19.5	9.08	28	12.85	39	18.17	55
360	6.71	18.6	9.49	26	13.42	37	18.97	53
390	6.98	17.9	9.87	25	13.96	36	19.75	51
430	7.33	17.1	10.37	24	14.66	34	20.74	48
470	7.66	16.3	10.84	23	15.33	33	21.68	46
510	7.98	15.7	11.29	22	15.97	31	22.58	44
560	8.37	14.9	11.83	21	16.73	30	23.66	42
620	8.80	14.2	12.45	20	17.61	28	24.90	40
680	9.22	13.6	13.04	19.2	18.44	27	26.08	38
750	9.68	12.9	13.69	18.3	19.37	26	27.39	37
820	10.12	12.4	14.32	17.5	20.25	25	28.64	35
910	10.67	11.7	15.08	16.6	21.33	23	30.17	33
1000	11.18	11.2	15.81	15.8	22.36	22	31.62	32

Table 3.8 **Maximum Permissible Voltage and Current for Preferred Value Resistors**

Note that values in the table for 270 ohms are not appropriate because 270 kΩ is not divisible by 270 ohms in multiples of 100.

Should powers in excess of 1 watt be under consideration, these are simply calculated by doubling the table values for ¼ of the power value required. The table is thus effective for powers up to 4 watts.

For example:
What is the maximum current permissible through a 330 ohm, 2 watt resistor?

¼ of 2 watts = ½ watt, so this is the appropriate table column to use.

The maximum current for 330 ohms (½ watt) is given as 39 mA, hence the value for 330 ohms (2 watts) = 39 x 2 = 78 mA.

EXAMPLE:
What are the smallest resistors (from the power rating aspect) which can be used in the circuit of Fig.3.6?

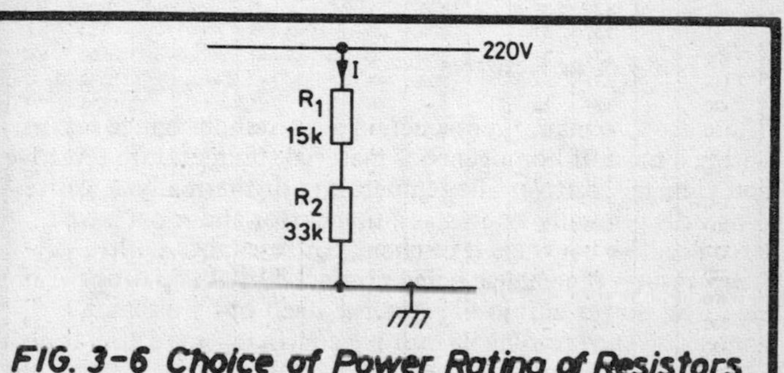

FIG. 3-6 Choice of Power Rating of Resistors

$$\text{Voltage across } R_1 = 220 \times \frac{15}{33 + 15} = 68.75\text{v.}$$

$$\text{Voltage across } R_2 = 220 - 68.75 = 151.25\text{v.}$$

Now, R_1 = 15 kΩ = 150 x 100 ohms

From table, maximum voltage for ¼ watt = 61.2
½ watt = 86.6

Thus a ½ watt resistor is the smallest usable

R_2 = 33 kΩ = 330 x 100 ohms

From table, maximum voltage for ¼ watt = 90.8
½ watt = 128.5
1 watt = 181.7

thus a 1 watt resistor is necessary.

Alternatively:

$$\text{Current } I = \frac{220}{48000} \times 1000 \text{ mA} = 4.58 \text{ mA}$$

R_1 – From table, maximum current for ¼ watt = 4.1 m
½ watt = 5.8 m
R_2 – From table, maximum current for ¼ watt = 2.8 m
½ watt = 3.9 m
1 watt = 5.5 m
giving the same requirements as above.

3.2.6 Non-Linear Resistors

There are two main types wherein a resistance change occurs within a mass of homogeneous material, temperature sensitive and voltage sensitive. The temperature or thermally sensitive resistor is generally known as a thermistor and most have a large negative coefficient of change of resistance with temperature, the change being produced by dissipating electrical power within the material itself or by a separate heater coil surrounding it. Positive coefficients are also availabl

A typical relationship between applied voltage and the current which flows in a directly heated (negative coefficient) thermistor is given in Fig.3.7. As V is increased from zero, the straightness of the characteristic shows that Ohm's Law is bein obeyed until a value marked as V_{max} is reached. This is because the heat is being dissipated as fast as it is being generated withi the material, the temperature of which therefore does not rise. At V_{max} conditions change radically because the heat generate is sufficiently great that not all is lost and the material temperature rises with consequent fall in resistance, ie the negative incremental resistance characteristic is obtained. The two curves show that as ambient temperature falls, V_{max} increases because heat lost from the material is greater.

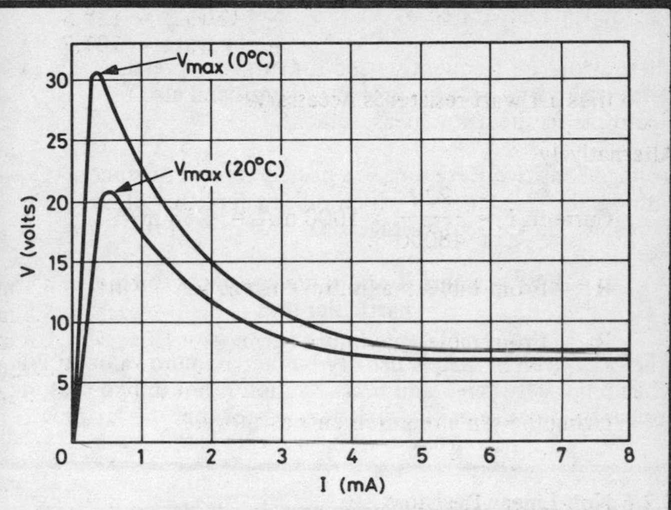

FIG. 3-7 Current/Voltage relationship for Thermistor

Generally for a thermistor, the resistance at temperature t°K follows approximately the law

$$R = Ae^{B/t} \text{ ohms}$$

where A and B are constants for the particular type of thermistor.

From the foregoing it is clear that the thermistor will not act until V_{max} is reached and held, albeit for a very short period of time. An approximate relationship from which V_{max} can be estimated is

$$V_{max} = \frac{\sqrt{R_o}}{K}$$

where R_0 is the "no current' resistance at the temperature under consideration
K is a constant for any particular type of thermistor

For the type illustrated by Fig.3.7, K = 20.

Thermistors are frequently used for surge suppression, e.g. in series with valve heaters, projector lamps and also for electronic temperature measurement.

Voltage-Sensitive Resistors, are usually a form of silicon carbide, the resistance of which falls as applied voltage increases. The current/voltage relationship is of the form

$I = KV^n$ where K and n are constants for the particular unit

When I is given in mA, K usually has a maximum value of 2-3 while n lies between 2 and 6. The action is not unlike that of copper oxide or selenium rectifiers except that the latter are

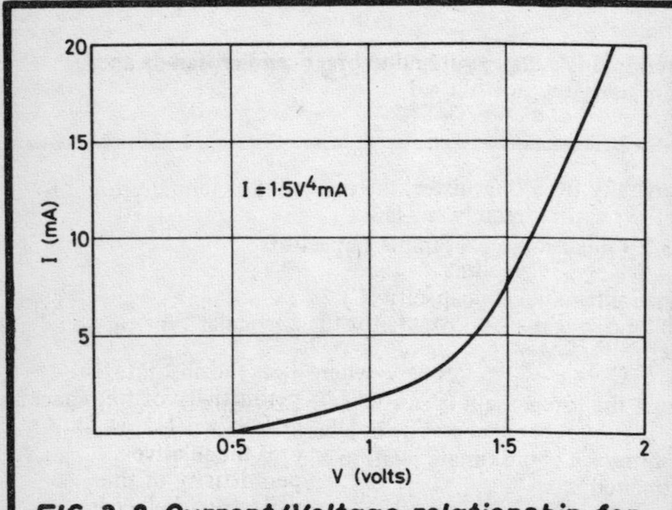

FIG. 3-8 Current/Voltage relationship for Voltage-Sensitive Resistor

capable of higher values of n. The use is as surge limiters, voltage stabilizers, spark quenches, loudness limiters on ear-phones etc. A typical characteristic is given in Fig.3.8. The capability of large resistance change can be judged from the fact that in this particular case, the resistance at 0.2V is about 80,000 ohms, whereas at 2.0V it has fallen to about 80 ohms.

3.3 CAPACITORS

3.3.1 Definition of Capacitance

For a given capacitor, the ratio of the charge Q (coulombs) to the p.d. (volts) is constant and this ratio is known as the capacitance (C) in farads.

$$\text{i.e. } C\,(\text{farads}) = \frac{Q\,(\text{coulombs})}{V\,(\text{volts})}$$

$$\text{also } Q = CV \text{ coulombs}$$

$$V = \frac{Q}{C} \text{ volts}$$

The farad is a large unit and in practice microfarads and picofarads are mostly used,

$$1\mu F = 10^{-6} \text{ farad}$$

$$1pF = 10^{-12} \text{ farad}$$

3.3.2 Construction of Simple Capacitors

General formula for capacity (C)

$$C = \epsilon_0 \epsilon_r \frac{A}{t} \text{ farads}$$

where ϵ_0 is the absolute permittivity of free space ($= 8.85 \times 10^{-12}$)
ϵ_r is the relative permittivity of the dielectric (dielectric constant)
A = cross-sectional area of dielectric in square metres
t = thickness of dielectric in metres.

Modifying this formula for more practical units and considering 2 plates only, each of A sq. cms area separated by air, since ϵ_r for air $\simeq 1$ (actually 1.00059). Then

(1) $$C = 0.0885 \frac{A}{t} \text{ picofarads}$$ where t is the separation between the plates in cms (ie dielectric thickness)

If a dielectric other than air is used, the capacitance above is multiplied by the value of ϵ_r for the dielectric ie

(2) $$C = 0.0885 \frac{A}{t} \epsilon_r \text{ picofarads}$$

Typical values of ϵ_r are

Ceramic	100–1000	Glass	4–8
Mica	6–7	Polystyrene	2–3
Shellac	2–4	Waxed Paper	5

If several plates are interleaved, the effective dielectric area is obtained by multiplying the effective area of one plate (ie excluding overlaps) by the number of dielectric spaces as in Fig.3.9 where in fact there are 4.

Then:

(3) $$C = 0.0885 \frac{A}{t} \epsilon_r N \text{ picofarads}$$ where N = number of dielectric spaces

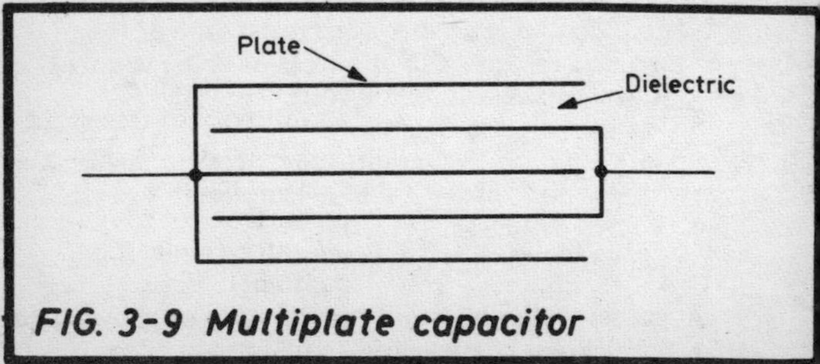

FIG. 3-9 Multiplate capacitor

EXAMPLE:

(1) A capacitor of 5 plates as in Fig.3.9, each 4 sq. cms. in area has a dielectric of glass having a relative permitivity of 5 and thickness 0.2 cms. What is the capacitance?

From formula (3) $C = 0.0885 \times \frac{4}{0.2} \times 5 \times 4 \text{ pf} = 35.4 \text{ pf}$

(2) A small, two-plate air-dielectric capacitor of 20 pf has to be constructed. The plates are to be separated by small diameter insulated washers 1 mm thick. What plate area is required?

From formula (1) $20 = 0.0885 \frac{A}{0.1}$ sq.cms.

$$\therefore A = 22.6 \text{ sq.cms.}$$

Many small inaccuracies inevitably creep in, for example the knowledge of the exact permittivity of a dielectric (other than air). Nevertheless the formulae do give a good starting point and as in Example 2, a quick means of estimating plate size etc.

3.3.3 Coding and Preferred Values

The colour coding system used to indicate the capacitance, tolerance and if required, voltage rating of small capacitors is identical to that used for resistors, hence Table 3.3 applies.

The three colours in the colour code representing the 1st Digit, 2nd Digit and Multiplier give the capacitance in picofarads.

A fourth band or spot shows the tolerance and if voltage rating is quoted (e.g. on moulded paper capacitors), one or two bands or spots give the rating in hundreds of volts, ie only one band or spot is necessary up to 900 volts.

The Preferred Values for the various tolerances are identical with those for resistors, hence Table 3.4 applies.

(Also see Appendix for other marking codes.)

3.3.4 Series Combinations

Consider Fig.3.10 in which two capacitors C_1 and C_2 are connected in series across a voltage V with voltages V_1 and V_2 existing across C_1 and C_2 as shown:

Then $V = V_1 + V_2$ and since the charges are equal, say Q,

$$V_1 = \frac{Q}{C_1} \quad \text{and} \quad V_2 = \frac{Q}{C_2}$$

$$\therefore \frac{V}{Q} = \frac{1}{C_1} + \frac{1}{C_2}$$

$$\therefore \frac{1}{C} = \frac{1}{C_1} + \frac{1}{C_2}$$ where C is the combined capacitance.

and it can similarly be shown that

$$\frac{1}{C} = \frac{1}{C_1} + \frac{1}{C_2} + \frac{1}{C_3} + - - - -$$

For two capacitors only in series $C = \dfrac{C_1 C_2}{C_1 + C_2}$

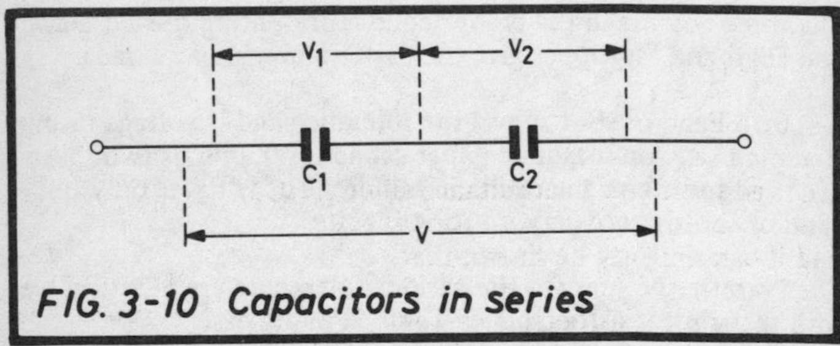

FIG. 3-10 Capacitors in series

These formulae have the same form as those for resistors in *parallel*, therefore Tables 3.6 and 3.7 apply except that whereas for resistors the table values are in ohms or multiples of ohms, for capacitors the values are in picofarads or microfarads or multiples i.e. the whole tables may be multiplied or divided by multiples of 10 as required.

3.3.5 Parallel Combinations

Fig.3.11 shows C_1 and C_2 connected in parallel. Each has the voltage V across it, producing charges Q_1 and Q_2.

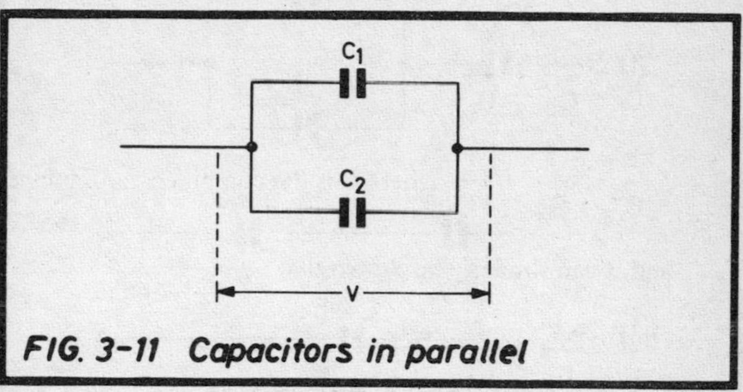

FIG. 3-11 Capacitors in parallel

Then, total charge $Q = Q_1 + Q_2$

Now $Q_1 = VC_1$ and $Q_2 = VC_2$

$\therefore Q = V(C_1 + C_2)$

$$\therefore \frac{Q}{V} = C_1 + C_2$$

and combined capacitance $C = C_1 + C_2$

and it can similarly be shown that

$$C = C_1 + C_2 + C_3 + - - - -$$

EXAMPLE:
What is the effective capacitance of the circuit in Fig.3.12(i)?

Let combined capacitance of C_2 and C_3 = Cp

$\therefore$ Cp = 2 + 4 = 6μF

(ii) Let C_1 in series with Cp = C

$$\text{Then } C = \frac{3 \times 6}{9} = 2\mu F \quad \ldots \ldots \ldots \ldots \text{(iii)}$$

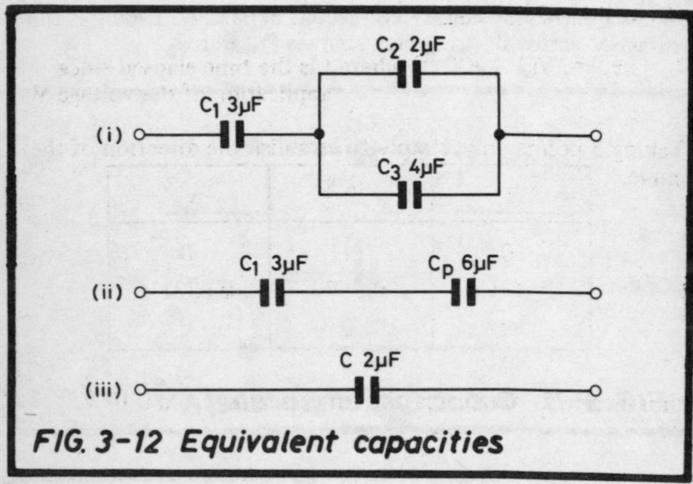

FIG. 3-12 Equivalent capacities

Capacitors in parallel are additive and hence analogous to resistors in series, thus Table 3.5 applies and again the whole table may be multiplied or divided by multiples of 10 as required.

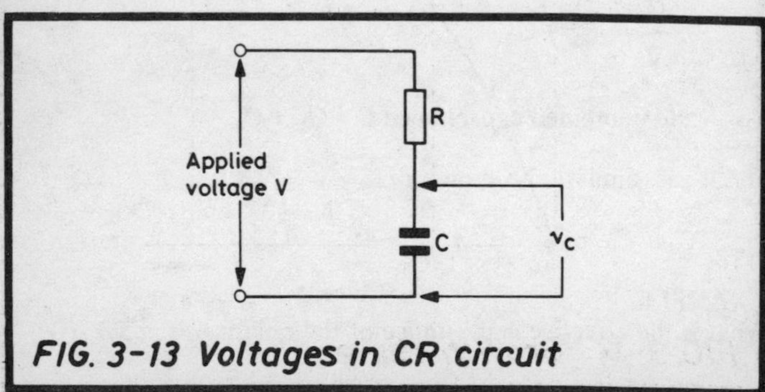

FIG. 3-13 Voltages in CR circuit

3.3.6 Time Constants

If a constant voltage V is maintained across a series combination of resistance R and capacitance C as in Fig.3.13, then the p.d. across the capacitance will take on an exponential form rising from 0 at the moment of application of V to the value V over a certain period of time as shown in Fig.3.14, the value of v_c at any instant being given by

$v_c = V(1 - e^{-t/CR})$ where t is the time elapsed since application of the voltage V

Taking 3 points only, simply to establish the direction of the curve,

t	$e^{-t/CR}$	v_c
0	1	0
CR	0.3679	0.6321V
∞	0	V

thus clearly as t increases, the curve is rising from 0 to V.

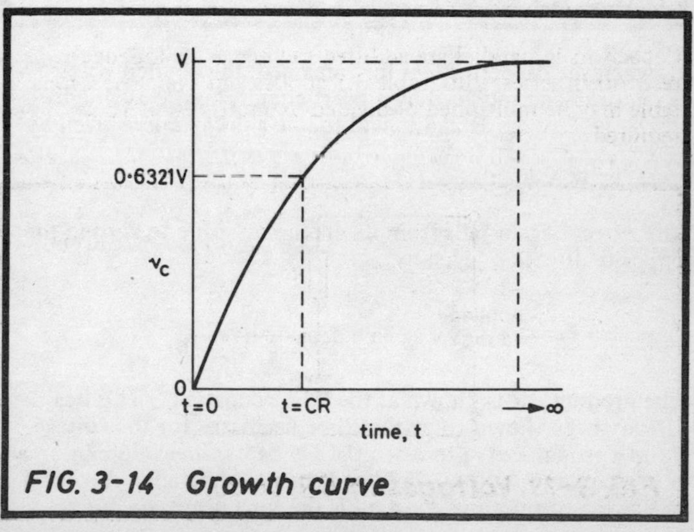

FIG. 3-14 Growth curve

The instantaneous current i in the circuit will in the same way follow a falling exponential curve and similarly it will be seen that at t = 0, CR and ∞, i = V/R, 0.3679 V/R and 0 respectively as in Fig.3.15.

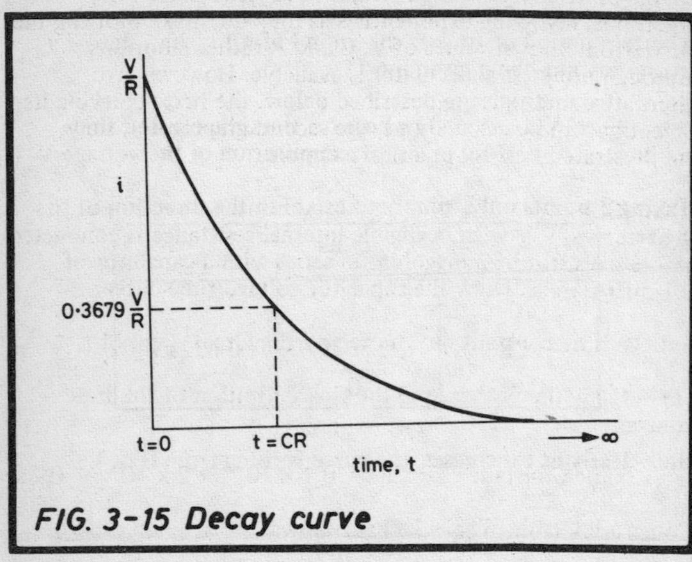

FIG. 3-15 Decay curve

If, when the capacitor C is fully charged, the applied voltage V is removed and replaced by a short-circuit, then C is able to discharge through R and v_c will follow a decay curve given by

$$v_c = V e^{-t/CR}$$

The current again falls from its maximum value to zero in the opposite direction given by

$$i = -\frac{V}{R} e^{-t/CR}$$, again a decay curve.

The product CR is known as the "time constant". This has already been shown to be that time necessary for the voltage v_c on a growth curve to reach 0.6321 of its final value, or on a decay curve to fall to 0.3679 of its maximum value. A second definition of the time constant is the time necessary for v_c

to attain its maximum value were it to continue increasing at the rate at which it is so doing at any particular instant. These definitions have considerable practical application when it is required to produce a CR circuit with a required time of charge or discharge to a particular value of voltage. The equations, involving exponentials as they do, make plotting the curve from known values of V, C and R rather laborious although not so if a calculator is available. However two alternative methods are described below, the first achieving its object by simple calculations, the second graphically, they are illustrated best by practical examples:

EXAMPLE:
A battery of 9 V with negligible internal resistance is connected across a resistor of 2 megohms in series with a capacitor of 0.1 microfarads. Draw the capacitor voltage/time curve.

Let the time constant be denoted by the Greek symbol τ.

(1) Using the principle of the first definition of the time constant

$$\text{Time Constant } \tau = CR = 0.1 \times 10^{-6} \times 2 \times 10^{6} = 0.2 \text{ secs.}$$

Construct a table (Table 3.9) as follows:

Each time-constant interval is treated separately, adding on to the previous value of v_c, (0.632 x the rise in voltage remaining). Thus Columns 2 and 3 are taken from Columns 5 and 6 of the previous line.

In practice, it is unlikely that the graph would be required beyond 0.6 secs because at this point the change in voltage with time is small and becomes progressively smaller as time increases. Fig.3.16 shows the graph of v_c (Column 5) plotted against time.

1	2	3	4	5	6
Time Interval (secs)	v_c at beginning of Interval	Voltage rise remaining at beginning of Interval	Voltage rise during Interval (= 0.632 x Col.3)	v_c at end of Interval (Cols. 2 + 4)	Voltage rise remaining at end of Interval
0 – 0.2	0	9	5.69	5.69	3.31
0.2 – 0.4	5.69	3.31	2.09	7.78	1.22
0.4 – 0.6	7.78	1.22	0.77	8.55	0.45
0.6 – 0.8	8.55	0.45	0.28	8.83	0.17
0.8 – 1.0	8.83	0.17	0.11	8.94	0.06

Table 3.9 **Calculations for Growth Curve**

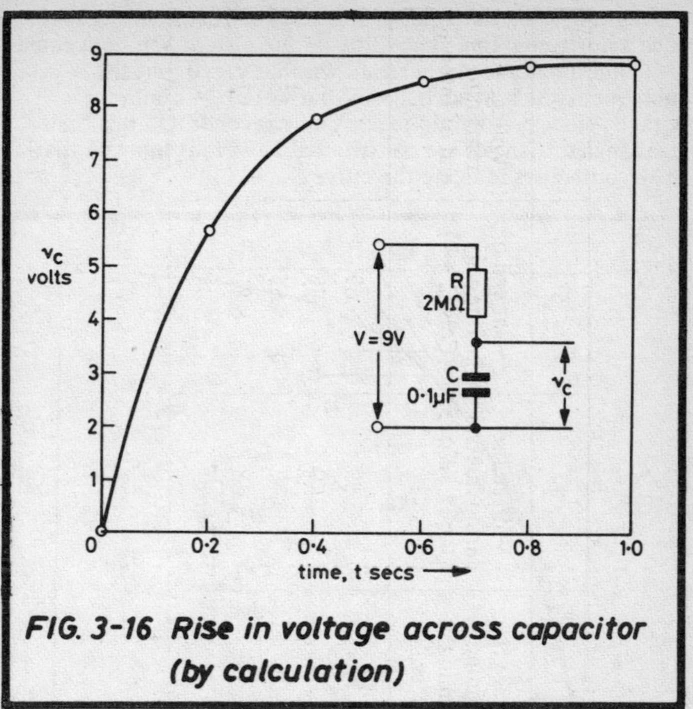

FIG. 3-16 Rise in voltage across capacitor (by calculation)

Table 3.9 shows the calculations in their entirety but in fact it is even more simple to express v_c as a fraction of V for each time-interval thus

t	v_c
0	0
τ	0.632 V
2τ	0.865 V
3τ	0.950 V
4τ	0.982 V

and in this particular example where V = 9v, v_c = 0.632 x 9, 0.865 x 9 etc, giving the same points on the curve of Fig.3.16.

(2) Using the principle of the second definition, a completely graphical method which is quick to construct is as shown in

Fig.3.17. The axes are drawn and labelled first as in Fig.3.16, then a horizontal line at V volts. At the instant V is connected, t = 0 and it has already been shown that $v_c = 0$, but the subsequent rise is at such a rate that were it to continue without change, it would reach V in τ seconds. On this basis right-angled triangles are constructed at various times so that the hypotenuses indicate the curve slopes.

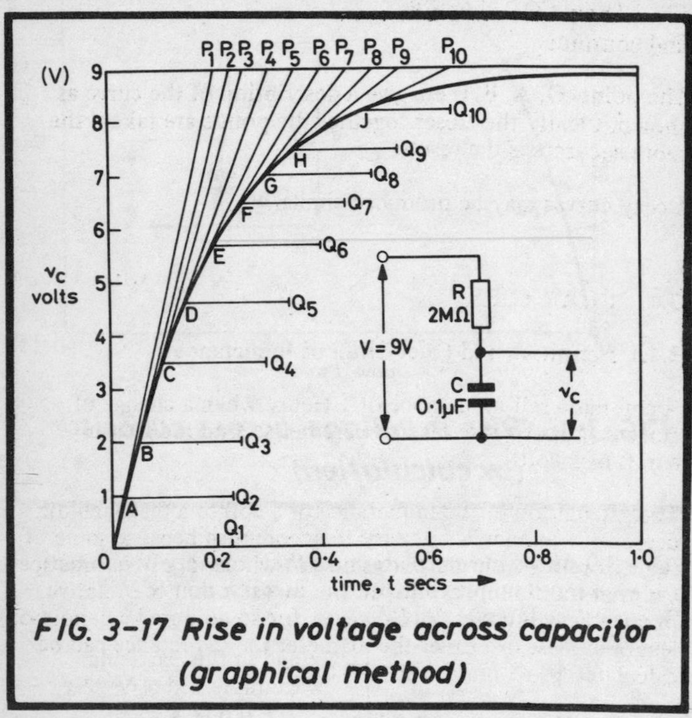

FIG. 3-17 Rise in voltage across capacitor (graphical method)

Note in Fig.3.17

OQ_1, AQ_2, BQ_3, CQ_4 — — — are all drawn to equal τ seconds.

Points, P_1 P_2 P_3 — — — are on the horizontal line drawn at V volts

P_1 is vertically above Q_1, P_2 above Q_2 etc.

(1) Draw OQ_1, then OP_1
(2) Mark point A at a short distance up OP_1
(3) Draw AQ_2, then AP_2
(4) Mark point B at a short distance up AP_2
(5) Draw BQ_3, then BP_3
(6) Mark point C at a short distance up BP_3
(7) Draw CQ_4, then CP_4
and continue.

The points O, A, B, C etc give a description of the curve as shown. Clearly the closer together the points are taken, the more accurate is the result.

Decay curves may be produced similarly.

3.4 INDUCTORS

3.4.1 Definition and Calculation of Inductance

A coil has a self-inductance of 1 Henry when a change of current in it at a rate of 1 ampere per second induces an e.m.f. of 1 volt.

The calculation of the inductance of a coil to a reasonable degree of accuracy is not easy to accomplish because some of the theoretical considerations do not wholly apply in practice e.g. that the flux links with all the turns or that the relative permeability is constant. However, for an air-cored solenoid of length at least five times the diameter the inductance can be calculated to within about 10% by

$$\text{Inductance, } L = \frac{\mu_o N^2 A}{l} \text{ henrys (H)}$$

where μ_o = permeability of free space ($4\pi \times 10^{-7}$ Henry/metre)
N = number of turns
A = cross sectional area of coil in sq. metres
l = length of coil in metres.

EXAMPLE:
What is the approximate inductance of an air-cored solenoid 12 cm. long and 2 cm. diameter, close wound with 2000 turns of wire?

$$\text{Inductance, } L = \frac{\mu_o N^2 A}{l} = \frac{4\pi \times 10^{-7} \times 4 \times 10^6 \times \pi \times 10^{-4}}{12 \times 10^{-2}}$$

$$= \frac{4\pi^2}{3} \times 10^{-3}\ \text{H} = 13.16\ \text{mH}$$

When a core is added calculation becomes more difficult because the relative permeability μ_r of the core varies with flux density, and especially difficult if the core is not closed, ie there is a relatively long air-path within the magnetic circuit. Nevertheless, for closed cores, some idea of the inductance can be obtained by extending the above formula with an estimation of the mean relative permeability thus

$$L = \frac{\mu_o N^2 A}{l} \times \mu_r$$

where μ_r is the mean relative permeability
l is the mean core path.

3.4.2 Construction of Air-Cored Inductors

Resistors and capacitors usually abound in plenty in the workshop, both new and second-hand and it is often useful to be able to make up a particular value from two or more others. Inductors, on the other hand seldom exist in large numbers conveniently colour-coded, they are usually made to order for a specific purpose. In this section therefore, more emphasis is placed on home construction although it is impossible to treat the subject at length because this would need a complete book on its own and in fact one has already been published in this series (No.160 "Coil Design and Construction Manual").

To allow for the inevitable small inaccuracies which creep into home construction it is a good idea to wind on a few more turns than are calculated to be necessary and then unwind until the desired results are obtained. Measurement of inductance is discussed in Section 6.2.3.

Most of the inaccuracy of the example in the preceding section arises from the fact that the flux due to the turns at the end of the coil does not cut as many other turns as happens at the centre, thus the self-inductance contribution is lower. Shape of coil therefore affects inductance considerably and it is necessary to build a factor into the formulae for this.

Much effort has already been expended in producing formulae for circular cross-section coils and of the several ones which exist, two are worthy of mention here:

(1) $$L = \frac{0.2\,N^2 d^2}{3.5d + 8l}\ \mu H$$

where L = inductance
N = number of turns
d = diameter of coil (strictly to centre of wire) in inches
l = length of winding in inches

In metric terms this becomes:

$$L = \frac{0.0787\,N^2\,d^2}{3.5d + 8l}\ \mu H$$

where d and l are in centimetres

The second published formula is

(2) $$L = \frac{r^2 \times N^2}{9r + 10l}\ \mu H$$

where r = outside radius of coil in inches
l = length of winding in inches

In metric terms and substituting d for r this becomes:

$$L = \frac{0.394\,N^2\,d^2}{18d + 40l}\ \mu H$$ where d and l are in centimetres.

It will be observed that these two formulae differ only slightly in the coefficient of d in the denominator.

An interesting formula which is offered here is one which takes greater account of the fact that the effect of the ratio l/d on the inductance follows approximately an exponential law:

(1) If l/d is within the range 0.3 – 1.0

$$L = \frac{0.007\ N^2\ d^{1.57}}{l^{0.57}}\ \mu H$$

where d and l are in centimetres

$$\therefore N = \sqrt{\frac{143\ L \times l^{0.57}}{d^{1.57}}}$$

where L is in microhenrys

(2) If l/d is within the range 1.2 – 8.0

$$L = \frac{0.0073\ N^2\ d^{1.863}}{l^{0.863}}\ \mu H$$

again where d and l are in centimetres

$$\therefore N = \sqrt{\frac{137\ L \times l^{0.863}}{d^{1.863}}}$$

where L is in microhenrys

EXAMPLE:
A coil of inductance 200μH is required, preferably on a circular former of 1.5 cm diameter and about 5 cms length. If such a coil is possible, how should it be wound?

$$l/d = \frac{5}{1.5} = 3.33, \text{ therefore formula (2) applies.}$$

$$N = \sqrt{\frac{137 \times 200 \times 5^{0.863}}{1.5^{1.863}}}$$

$$\therefore N = \sqrt{\frac{137 \times 200 \times 4.011}{2.129}} = 227$$

$$\log 5 = 0.6990$$
$$0.863 \log 5 = 0.6032$$
$$\therefore 5^{0.863} = 4.011$$

$$\log 1.5 = 0.1761$$
$$1.863 \log 1.5 = 0.3281$$
$$\therefore 1.5^{1.863} = 2.129$$

Nominal Bare diam. mm	TURNS/cm (min)			Nominal Bare diam. mm	TURNS/cm (min)		
	Grade 1	Grade 2	Grade 3		Grade 1	Grade 2	Grade 3
2.000	4.8	4.7	4.6	0.500	18.3	17.6	16.9
1.900	5.0	5.0	4.9	0.450	20.2	19.4	18.6
1.800	5.3	5.2	5.1	0.400	22.6	21.7	20.7
1.700	5.6	5.5	5.4	0.355	25.3	24.2	23.0
1.600	5.9	5.8	5.7	0.315	28.4	27.0	25.6
1.500	6.3	6.2	6.1	0.280	31.8	29.9	28.3
1.400	6.8	6.6	6.5	0.250	35.2	33.2	31.3
1.320	7.2	7.0	6.9	0.224	39.1	36.8	34.5
1.250	7.5	7.4	7.2	0.200	43.5	40.8	38.3
1.180	8.0	7.8	7.6	0.180	47.9	45.1	42.2
1.120	8.4	8.2	8.0	0.160	53.5	50.3	47.0
1.060	8.8	8.7	8.5	0.140	60.2	56.8	52.9
1.000	9.4	9.1	8.9	0.125	67.1	62.9	58.5
0.950	9.9	9.6	9.3	0.112	74.6	69.9	64.5
0.900	10.4	10.1	9.8	0.100	82.6	77.5	70.9
0.850	11.0	10.7	10.4	0.090	90.9	85.5	78.1
0.800	11.6	11.3	11.0	0.080	102.0	95.2	86.2
0.750	12.4	12.0	11.7	0.071	113.6	105.3	ϕ
0.710	13.0	12.7	12.3	0.063	128.2	117.6	ϕ
0.630	14.6	14.2	13.7	0.050	161.3	147.1	ϕ
0.560	16.4	15.8	15.2				

Table 3.10 Turns Per Centimetre for Enamelled Copper Wires*

* Oleo - Resinous and Synthetic

ϕ not available

If enamelled wire is used and the winding itself is 5 cm long, then from Table 3.10 a suitable wire would be 0.180 mm which with Grade 1 insulation winds to 47.9 turns per cm, thus 227 turns would occupy 4.74 cms.

3.4.3 Magnetic Cored Inductors

The theory is simple, i.e. the inductance of a coil with a magnetic core is the inductance of the same coil with an air core multiplied by the relative permeability of the magnetic core. In practice however, the calculation of self-inductance with any pretence to accuracy is difficult because the core relative permeability is not constant, depending as it does upon the flux density which itself depends upon the current, i.e. the current/flux relationship is not linear. Account may also be taken of a small air-gap in the core but this truly becomes difficult if the air-gap is as long as the core itself, i.e. an open core as for example, with coils on a ferrite rod. Nevertheless, provided that a figure for the mean relative permeability of the complete magnetic circuit can be obtained, then the approximate inductance is calculated as above.

To take an air-gap into account it is necessary to calculate the total reluctance of the complete magnetic circuit by adding together the individual reluctances of the air-gap and the magnetic core on the basis that the reluctance (S) of a single homogeneous path is given by:

$$S = \frac{l}{\mu a}$$

where l = length of path
a = cross-sectional area of path
μ = permeability of medium (= 0 for air)

and the total reluctance

$$S_T = S_1 + S_2 + -- \quad = \frac{l_1}{\mu_1 a_1} + \frac{l_2}{\mu_2 a_2} + ---$$

therefore because inductance is inversely proportional to reluctance, comparison may be made of the inductance of a coil and magnetic core with or without air-gap(s). A further cause of inaccuracy with these formulae is that not all the flux produced within the magnetic core flows directly across the air-gap.

Nevertheless the formulae do give a starting point which can then be followed by "cut and try" bench work to obtain the desired results.

3.4.4 Series and Parallel Combinations

Combinations of inductances in series or parallel are less likely to be needed in the workshop because inductors are usually wound as required. However, the combination formulae are easily expressed because they are of identical form to those for resistances e.g.

Series $L = L_1 + L_2 + L_3 + ---$

where L is the combined inductance of the individual inductances L_1, L_2, L_3 etc.

Parallel $$\frac{1}{L} = \frac{1}{L_1} + \frac{1}{L_2} + \frac{1}{L_3} + ---$$

and for two inductances only in parallel

$$L = \frac{L_1 L_2}{L_1 + L_2}$$

and therefore, although not so labelled, Tables 3.5, 3.6 and 3.7 apply fully by simply substituting L for R whether L is in microhenrys, millihenrys or henrys.

These formulae are valid only when the individual inductors have no mutual coupling, i.e. the magnetic field of one does not cut the turns of another. When this does happen, Section 3.4.6 applies.

3.4.5 Time Constants

If a constant voltage V, is applied to a series combination of resistance R and inductance L as in Fig.3.18, then the current i will take on an exponential form rising from 0 at the moment of application of V to the maximum value of V/R over a certain period of time, the curve shape being as shown by Fig.3.14 for capacitors.

The instantaneous value of i for a time t elapsed is

$$i = \frac{V}{R}(1 - e^{-Rt/L}) = I(1 - e^{-Rt/L})$$ where I is the final steady value of current.

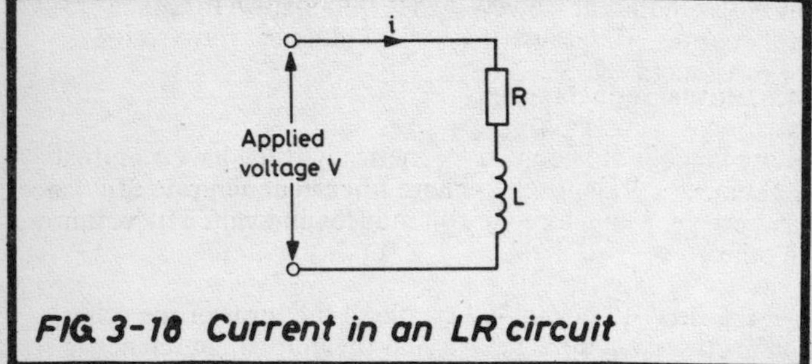

FIG. 3-18 Current in an LR circuit

The similarity of the build-up of current in an inductance with the build-up of voltage across a capacitor is clear, many of the principles developed in Section 3.3.6, therefore apply and are not duplicated in this section.

The voltage across the resistance R, (v_R) = i x R and therefore follows the same exponential law i.e. it too follows a growth curve.

However, since the voltage (v_L) across the inductance L is equal to $V - v_R$. then as v_R increases, v_L falls and in fact

$$v_L = V\,e^{-Rt/L}$$ giving a decay curve as shown in Fig.3.15.

In this case, the expression R/L is known as the time constant (τ). With R in ohms, L in henrys, τ is expressed in seconds.

Curves such as in Fig.3.16 and 3.17 may be drawn for any LR circuit using $\tau = L/R$.

If, after steady state conditions have been reached, ie the current i has reached the maximum I, the voltage V is removed and replaced by a short-circuit, the magnetic field collapses and in so doing drives a current back through R and the short-circuit, the current falling as the energy remaining in the magnetic field decreases, giving now a decay curve of the form

$$i = I e^{-Rt/L}$$

and similarly for the voltage across the resistance, v_R.

3.4.6 Mutual Inductance

When two coils are coupled magnetically, they have a mutual inductance of 1 Henry if a change of current at a rate of 1 ampere per second in one coil induces an e.m.f. of 1 volt in the other.

If all the flux of one coil links with all the turns of the other then maximum coupling exists and the mutual inductance M is given by

$$M^2 = L_1 L_2 \qquad \therefore M = \sqrt{L_1 L_2}$$

where L_1 and L_2 are the self-inductances of the two coils.

This condition is unlikely to be achieved in practice and a coupling coefficient k is employed to denote the degree of coupling, it is obtained from the ratio of M to $\sqrt{L_1 L_2}$ ie

$$M = k\sqrt{L_1 L_2}$$

With two coils coupled as closely as possible, k can reach a value of 0.98 – 0.99 while when the coils are completely separated, k = 0. In radio air-cored coils, k is usually less than 0.5.

When the two coils L_1 and L_2 are themselves connected in series, the combination inductance value is $L_1 + L_2 \pm 2M$ depending on whether they are in series – aiding or opposition. The term 2M arises from the fact that both coils are affected by flux linkages with the other coil. Use is made of this feature in the measurement of mutual inductance (see Section 6.2.3).

3.5 TRANSFORMERS

3.5.1 General Theory

Transformer action arises from the fact that the magnetic flux produced by the current in one coil links with the turns in the other coil(s). The Mutual Inductance which therefore exists between the coils is defined in Section 3.4.6.

For a sinusoidal current I_p in the primary winding, the e.m.f. generated in the secondary is

$$e_s = \pm j\omega MI_p \qquad \text{or } \omega MI_p \ \angle \pm 90^\circ$$

the angle being positive or negative according to the sense in which the windings are connected.

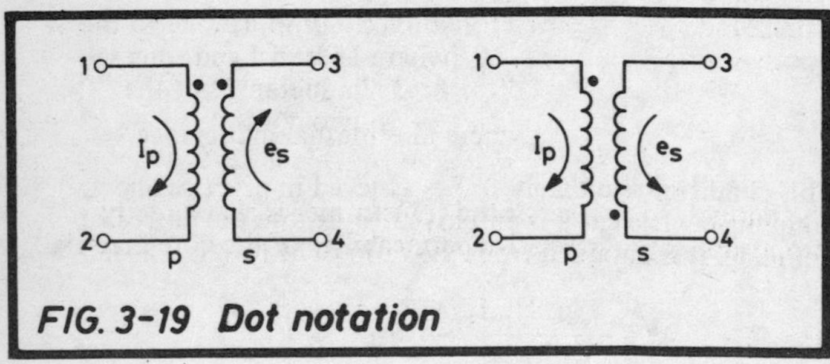

FIG. 3-19 Dot notation

To avoid confusion, the dot notation is frequently used on diagrams as shown in Fig.3.19. The dots are placed at the ends of the winding symbols to indicate that currents entering at those ends product magnetic fluxes in the core in the same direction.

The basic principles of coupled circuits are considered in greater depth in Section 4.7.5 which deals with h.f., low coupling factor transformers. These principles apply equally to the more generally recognised transformer, i.e. iron-cored, lower frequency and higher coupling factor (approaching unity).

Transformers with Iron Cores

By including iron or an alloy in the magnetic path of a transformer, the flux is greatly increased because of the increase in relative permeability. The expression for the primary inductance is

$$L_P = \frac{\mu_o \mu_r N_p^2 A}{l} \text{ henry}$$

where μ_o = permeability of free space
μ_r = relative permeability of the core material
N_p = no. of turns on primary
A = cross-sectional area of coil in sq. m.
l = length of coil in metres

and similarly for the inductance, L_s of the secondary winding.

The inductances are therefore directly proportional to the relative permeability, μ_r, of the core material and since

$$M = k\sqrt{L_p L_s}$$ where M = mutual inductance

the mutual inductance of the transformer is also directly proportional to the relative permeability of the core material.

The flux $$\Phi = \frac{\mu_o \mu_r N_p I_p A}{l} \text{ Wb}$$

and the r.m.s. value of the induced secondary voltage,

$$e_s = 4.44 \, N \, \Phi_{max} \, f \text{ volts}$$ where f is the frequency of the current I_p in Hz.

Voltage Ratio

Considering an ideal transformer, since, as above a flux Φ_{max} produces a voltage e_s in the secondary proportional to the secondary turns, then because the same flux cuts the primary winding, the primary induced voltage is proportional to the primary turns i.e.

$$\frac{e_s}{e_p} = \frac{4.44\, N_s\, \Phi_{max}\, f}{4.44\, N_p\, \Phi_{max}\, f} \quad \therefore \quad \frac{e_s}{e_p} = \frac{N_s}{N_p} \quad \text{or equally}$$

$$\frac{V_s}{V_p} = \frac{N_s}{N_p}$$

thus the secondary voltage is $\frac{N_s}{N_p}$ times the primary voltage, that is

the voltage ratio is equal to the turns ratio.

Current Ratio

Again considering an ideal transformer, when its secondary is connected to a resistive load (R_s) the effective primary impedance becomes mainly a resistive component $(\omega M)^2/R_s$ which has no phase angle.

Since output power = input power (assuming no losses)

$$V_p I_p = V_s I_s \quad \text{and since} \quad \frac{V_p}{V_s} = \frac{N_p}{N_s}$$

$$\frac{I_s}{I_p} = \frac{N_p}{N_s} \qquad \text{that is}$$

the current ratio is inversely proportional to the turns ratio.

3.5.2 Impedance Matching

As above, for the ideal transformer

$$V_pI_p = V_sI_s \quad \text{and} \quad Ip = \frac{V_p}{R_p}, \; I_s = \frac{V_s}{R_s}$$

where R_p and R_s are the primary and secondary load resistances.

$$\therefore \frac{V_p^2}{R_p} = \frac{V_s^2}{R_s}$$

$$\therefore \frac{R_s}{R_p} = \left(\frac{N_s}{N_p}\right)^2 \quad \text{that is}$$

the resistance ratio is equal to the (turns ratio)^2

By similar reasoning it can be shown that if Z_p and Z_s are the appropriate impedances

$$\frac{Z_s}{Z_p} = \left(\frac{N_s}{N_p}\right)^2 \quad \text{that is}$$

the impedance ratio is also equal to the (turns ratio)^2

A common use for transformers in electronic systems is for matching a load to a source to obtain maximum power transfer (Section 5.1.5). A typical example is the matching of loudspeakers to power amplifiers (although transistor power amplifiers are designed wherever possible to avoid the bulk and expense of output transformers) e.g. an output stage requiring an optimum load of 2000 ohms and feeding a loudspeaker of impedance 8 ohms. The transformer needed in this instance must have a turns ratio of

$$\sqrt{\frac{2000}{8}} = 15.8 : 1$$

↓Ω SOURCE \ LOAD → Ω	1		2		3		4		5		6		7		8		9		10	
	A	B	A	B	A	B	A	B	A	B	A	B	A	B	A	B	A	B	A	B
1	1.000	3.162	.7071	2.236	.5773	1.826	.5000	1.581	.4472	1.414	.4083	1.291	.3780	1.195	.3535	1.118	.3333	1.054	.3162	1.000
2	1.414	4.472	1.000	3.162	.8165	2.582	.7071	2.236	.6325	2.000	.5774	1.826	.5345	1.690	.5000	1.581	.4714	1.491	.4472	1.414
3	1.732	5.477	1.225	3.873	1.000	3.162	.8660	2.739	.7746	2.450	.7071	2.236	.6547	2.070	.6124	1.936	.5774	1.826	.5477	1.732
4	2.000	6.325	1.414	4.472	1.155	3.652	1.000	3.162	.8944	2.828	.8165	2.582	.7559	2.390	.7071	2.236	.6667	2.108	.6325	2.000
5	2.236	7.071	1.581	5.000	1.291	4.083	1.118	3.535	1.000	3.162	.9129	2.887	.8452	2.673	.7906	2.500	.7454	2.357	.7071	2.236
6	2.450	7.746	1.732	5.477	1.414	4.472	1.225	3.873	1.095	3.464	1.000	3.162	.9258	2.928	.8660	2.739	.8165	2.582	.7746	2.450
7	2.646	8.367	1.871	5.916	1.528	4.830	1.323	4.183	1.183	3.742	1.080	3.416	1.000	3.162	.9354	2.958	.8819	2.789	.8367	2.646
8	2.828	8.944	2.000	6.325	1.633	5.164	1.414	4.472	1.265	4.000	1.155	3.652	1.069	3.380	1.000	3.162	.9428	2.981	.8944	2.828
9	3.000	9.487	2.121	6.708	1.732	5.477	1.500	4.743	1.342	4.243	1.225	3.873	1.134	3.586	1.061	3.354	1.000	3.162	.9487	3.000
10	3.162	10.00	2.236	7.071	1.826	5.773	1.581	5.000	1.414	4.472	1.291	4.083	1.195	3.780	1.118	3.535	1.054	3.333	1.000	3.162

MULTIPLY OR DIVIDE COLUMN A or B FIGURES AS SHOWN

↓Ω SOURCE \ LOAD → Ω	1 – 10		10 – 100		100 – 1000		1000 – 10,000		10,000 – 100,000	
	A	B	A	B	A	B	A	B	A	B
1 – 10	NIL			÷ 10	÷ 10			÷ 100	÷ 100	
10 – 100		NIL	NIL			÷ 10	÷ 10			÷ 100
100 – 1000	× 10			NIL	NIL			÷ 10	÷ 10	
1000 – 10,000		× 10	× 10			NIL	NIL			÷ 10
10,000 – 100,000	× 100			× 10	× 10			NIL	NIL	

(NIL indicates column to be used with no correction)

Table 3.11 Values of Turns Ratio $T = N_1/N_2$ for Impedance Ratio Source/Load

Table 3.11 makes this calculation quickly for a wide range of practical ratios of source and load values commencing with a single integer – if the actual values lie between the table values, the table is still useful as a guide or as a check on calculations. In the above example, the answer is given exactly by the table which is used as follows

(i) take the first figure of both source and load impedances and the table gives two ratios marked A and B. In the example against a source value of 2 ohms and load value of 8 ohms are ratios A, .5000 and B, 1.581.

(ii) reference to the lower section of the table shows that for a source in the 1000 – 10,000 ohm range, and load in the 1 – 10 ohm range, the B figure is used and multiplied by 10, giving an answer of 15.81.

However, if the source impedance happened to be 2400 ohms, then the correct answer must lie between 15.81 and 19.36 (for sources of 2000 and 3000 ohms respectively), a guess might put the value around 17.

EXAMPLE:
A generator of e.m.f. 20V and internal resistance 600 ohms is to be matched by means of a transformer to a 1000 ohm line. Assuming an ideal transformer, what turns ratio is required and what is the power in the load with and without transformer?

From Table 3.11 the turns ratio required is 0.7746 : 1.

Power into load with transformer – the circuit equivalent is that of a 600 ohm load (R_L) across the generator terminals. Let I_1 be the current,

$$\text{then } I_1 = \frac{20}{600 + 600} \text{ amps}$$

$$\text{Power into load} = I_1^2 R_L = \left(\frac{20}{1200}\right)^2 \times 600 = 167 \text{ mW}$$

Power into load without transformer Let I_2 be the current

$$\text{then } I_2 = \frac{20}{600 + 1000} \text{ amps}$$

$$\text{Power into load} = I_2^2 R_L = \left(\frac{20}{1600}\right)^2 \times 1000 = 156 \text{ mW}$$

so demonstrating the greater power in the load when it is matched to the source.

4. ALTERNATING CURRENT CIRCUITS

4.1 DERIVATION OF THE SINUSOIDAL WAVEFORM

Consider a simple alternator having two poles and one single loop of wire rotating on an axis OQ within the magnetic field as shown in Fig.4.1. Faraday's Law (Sect.2.3) shows that the e.m.f. induced in the sides AB and CD of the loop is proportional to the rate of change of flux through each wire, therefore when passing the centre of a pole the rate of change of flux is greatest and the e.m.f. induced is maximum, conversely when the wires are midway between the poles, their movement is in a plane parallel to the flux and the e.m.f. induced is zero. Fleming's Right Hand Rule (Sect.2.3) enables the direction of the e.m.f. to be determined. The sections of the loop, AD and BC contribute nothing because they too rotate in a plane parallel to the flux.

For the purpose of diagrammatic illustration, OA, half of AD is considered as a rotating vector as projected to the right of the alternator in the figure and to the right of this the waveform is developed of generator output voltage, or equally, if the external circuit is closed, the current in OA.

It will be seen that at θ° from position OA, the height of the wave is equal to PA_2 and

$$PA_2 = \sin\theta \times OA = V\sin\theta$$ where V is the maximum voltage.

Thus the waveform is considered sinusoidal because its height at any point is proportional to the sine of the angle through which it has turned.

A 360° rotation of OA produces one complete wave. For certain calculations radians ($360^\circ = 2\pi$ radians) or time scales are used. The time scale marked is the practical one for the public supply mains.

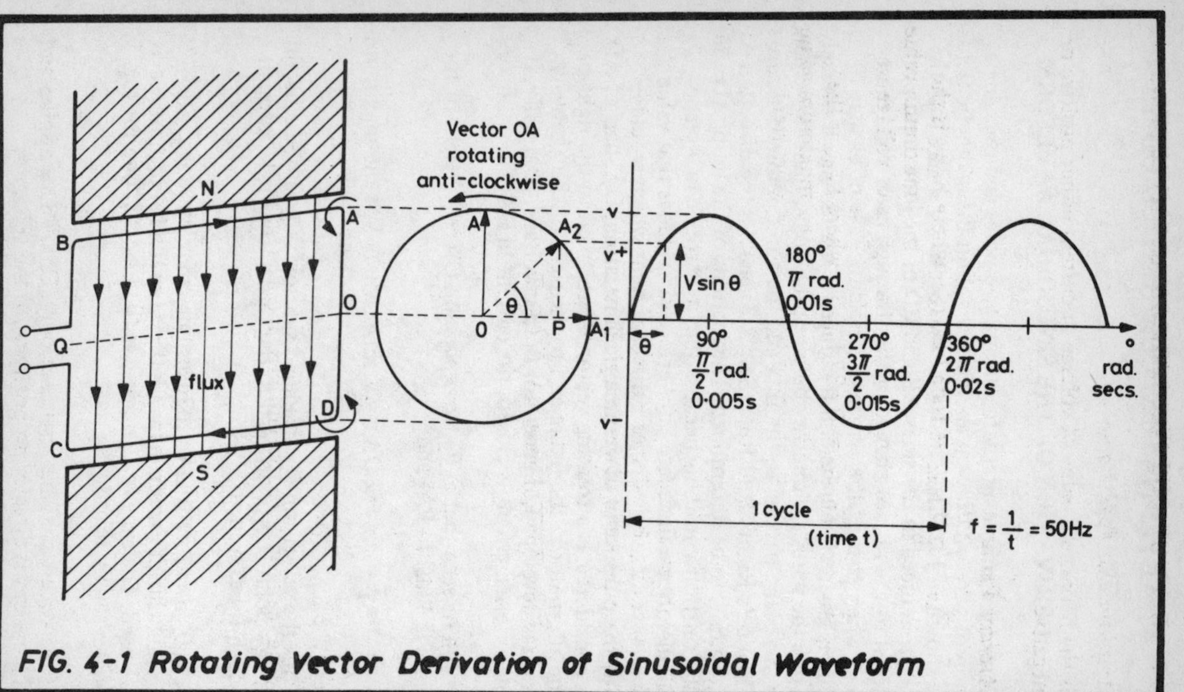

FIG. 4-1 Rotating Vector Derivation of Sinusoidal Waveform

4.2 DEFINITIONS FOR SINUSOIDAL WAVEFORMS

Amplitude or Peak Value

is the maximum value of voltage or current either positive or negative.

Average Value

$$E_{av} = \frac{2}{\pi} E_{max} = 0.637\, E_{max}$$ where E_{max} is the maximum value of voltage

$$I_{av} = \frac{2}{\pi} I_{max} = 0.637\, I_{max}$$ where I_{max} is the maximum value of current

RMS Value

The "root mean square" or "effective" value is that value which equals the value of a direct current which would dissipate the same power in a given resistor.

$$E_{rms} = \frac{1}{\sqrt{2}} E_{max} = 0.707\, E_{max}$$

$$I_{rms} = \frac{1}{\sqrt{2}} I_{max} = 0.707\, I_{max}$$

Form Factor

is the relationship between r.m.s. and average values

$$\therefore \text{Form Factor} = \frac{\frac{1}{\sqrt{2}}}{\frac{2}{\pi}} = \frac{\pi}{2\sqrt{2}} = 1.11$$

If the circuit contains resistance only, Ohm's Law applies and the current is directly proportional to the voltage.

4.3 REACTANCE

4.3.1 Capacitive Reactance

The charge on a capacitor is in phase with the p.d. applied to it as shown in Fig.4.2. The current is proportional to the rate of change of charge and therefore follows the curve shown, leading the voltage by 90°.

Reactance $X_C = \dfrac{1}{2\pi fC} = \dfrac{1}{\omega C}$ ohms where f = frequency in Hz
C = capacitance in farads
$\omega = 2\pi f$.

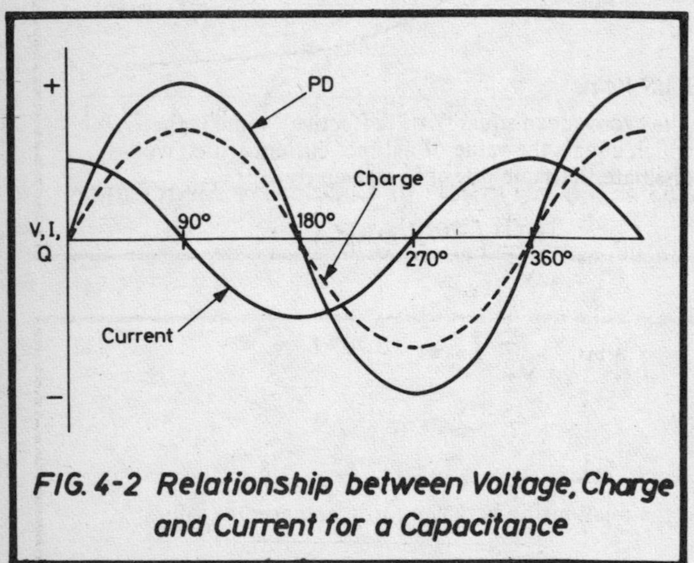

FIG. 4-2 Relationship between Voltage, Charge and Current for a Capacitance

The reactance falls with frequency as shown by Fig.4.3(i) and conventionally for impedance calculations is given a negative sign.

For calculations and diagrams the phase difference between voltage and current can be expressed by a vector, or phasor, diagram, see Fig.4.3(ii) where V_C represents a voltage applied to a capacitor and I_C represents the resultant current 90° ahead as also shown in Fig.4.2. Phasors are considered to rotate anti-clockwise.

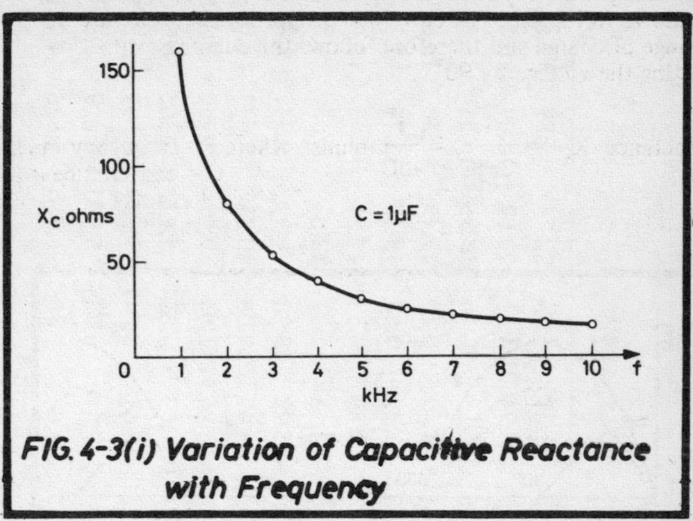

FIG. 4-3(i) Variation of Capacitive Reactance with Frequency

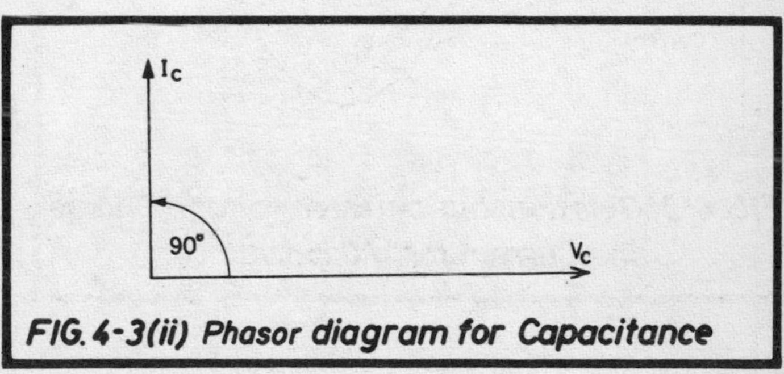

FIG. 4-3(ii) Phasor diagram for Capacitance

In electronic design an approximate idea of capacitive reactance values is frequently required, e.g. for by-pass or coupling capacitors. Table 4.1 is presented to enable a quick estimate to be made without the difficulty of reading fine lines on charts, nomographs etc. The table values are correct to four figures for the values and multiples of C and f as shown. Some

$C_{\mu F}$ \ f_{kHz}	1	2	3	4	5	6	7	8	9	10
1.0	159.2	79.58	53.05	39.79	31.83	26.53	22.74	19.89	17.68	15.92
1.1	144.7	72.34	48.23	36.17	28.94	24.11	20.67	18.09	16.08	14.47
1.2	132.6	66.31	44.21	33.16	26.53	22.10	18.95	16.58	14.74	13.26
1.3	122.4	61.21	40.81	30.61	24.49	20.40	17.49	15.30	13.60	12.24
1.5	106.1	53.05	35.37	26.53	21.22	17.68	15.16	13.26	11.79	10.61
1.6	99.47	49.74	33.16	24.87	19.89	16.58	14.21	12.43	11.05	9.947
1.8	88.42	44.21	29.47	22.10	17.68	14.74	12.63	11.05	9.824	8.842
2.0	79.58	39.79	26.53	19.89	15.92	13.26	11.37	9.947	8.842	7.958
2.2	72.34	36.17	24.11	18.09	14.47	12.06	10.33	9.043	8.038	7.234
2.4	66.31	33.16	22.10	16.58	13.26	11.05	9.474	8.289	7.368	6.631
2.7	58.95	29.47	19.65	14.74	11.79	9.824	8.421	7.368	6.550	5.895
3.0	53.05	26.53	17.68	13.26	10.61	8.842	7.579	6.631	5.895	5.305
3.3	48.23	24.11	16.08	12.06	9.646	8.038	6.890	6.029	5.359	4.823
3.6	44.21	22.10	14.74	11.05	8.842	7.368	6.316	5.526	4.912	4.421

3.9	40.81	20.49	13.60	10.20	8.162	6.801	5.830	5.101	4.534	4.081
4.0	39.79	19.89	13.26	9.947	7.958	6.631	5.684	4.974	4.421	3.979
4.3	37.01	18.51	12.34	9.253	7.403	6.169	5.288	4.627	4.113	3.701
4.7	33.86	16.93	11.29	8.466	6.773	5.644	4.838	4.233	3.763	3.386
5.0	31.83	15.92	10.61	7.958	6.366	5.305	4.547	3.979	3.537	3.183
5.1	31.21	15.60	10.40	7.802	6.241	5.201	4.458	3.900	3.467	3.121
5.6	28.42	14.21	9.474	7.105	5.684	4.737	4.060	3.553	3.158	2.842
6.0	26.53	13.26	8.842	6.631	5.305	4.421	3.789	3.316	2.947	2.653
6.2	25.67	12.84	8.557	6.418	5.134	4.278	3.667	3.209	2.852	2.567
6.8	23.41	11.70	7.802	5.851	4.681	3.901	3.344	2.926	2.601	2.341
7.0	22.74	11.37	7.579	5.684	4.547	3.789	3.248	2.842	2.526	2.274
7.5	21.22	10.61	7.074	5.305	4.244	3.537	3.032	2.653	2.358	2.122
8.0	19.89	9.947	6.631	4.974	3.979	3.316	2.842	2.487	2.210	1.989
8.2	19.41	9.705	6.470	4.852	3.882	3.235	2.773	2.426	2.157	1.941
9.0	17.68	8.842	5.895	4.421	3.537	2.947	2.526	2.210	1.965	1.768
9.1	17.49	8.745	5.830	4.372	3.498	2.915	2.499	2.186	1.943	1.749
10.0	15.92	7.958	5.305	3.979	3.183	2.653	2.274	1.989	1.768	1.592

Multiply Table Figures for Ranges of f and C as shown:

f → / C ↓	10-100 Hz	100-1000 Hz	1-10 kHz	10-100 kHz	100 kHz -1 MHz	1-10 MHz	10-100 MHz	100 MHz -1 GHz
1-10 pF	$\times 10^8$	$\times 10^7$	$\times 10^6$	$\times 10^5$	$\times 10^4$	$\times 10^3$	$\times 10^2$	$\times 10$
10-100 pF	$\times 10^7$	$\times 10^6$	$\times 10^5$	$\times 10^4$	$\times 10^3$	$\times 10^2$	$\times 10$	NIL
100-1000 pF	$\times 10^6$	$\times 10^5$	$\times 10^4$	$\times 10^3$	$\times 10^2$	$\times 10$	NIL	$\div 10$
.001-.01 μF	$\times 10^5$	$\times 10^4$	$\times 10^3$	$\times 10^2$	$\times 10$	NIL	$\div 10$	$\div 10^2$
.01-0.1 μF	$\times 10^4$	$\times 10^3$	$\times 10^2$	$\times 10$	NIL	$\div 10$	$\div 10^2$	$\div 10^3$
0.1-1.0 μF	$\times 10^3$	$\times 10^2$	$\times 10$	NIL	$\div 10$	$\div 10^2$	$\div 10^3$	$\div 10^4$
1-10 μF	$\times 10^2$	$\times 10$	NIL	$\div 10$	$\div 10^2$	$\div 10^3$	$\div 10^4$	$\div 10^5$
10-100 μF	$\times 10$	NIL	$\div 10$	$\div 10^2$	$\div 10^3$	$\div 10^4$	$\div 10^5$	$\div 10^6$

NIL – indicates no correction

Table 4.1 **Capacitive Reactance Values (ohms)**

approximation arises for other values which are obtained by interpolation.

EXAMPLE:
What is the reactance of a 68pF capacitor at 6 MHz?

From Table 4.1 at C = 6.8 μF, f = 6 kHz, reactance = 3.901 ohms

For C in the 10-100 pF range,
" f " " 1-10 MHz " multiply by 10^2

$\therefore$ Reactance = 3.901×10^2 = **390.1** ohms.

EXAMPLE:
What is the reactance of a 68pF capacitor at 6.5 MHz?

In this case interpolation is necessary:

At 6 MHz reactance is 390.1 ohms (as above)
At 7 MHz " " 334.4 "

6.5 MHz is half way between 6 and 7 MHz. 362 ohms is half way between 334 and 390 ohms $\therefore$ Answer **362 ohms**

(This, in fact, is only 2 ohms, about 0.5%, inaccurate)

4.3.2 Inductive Reactance

The 'back e.m.f.' set up in an inductor by a changing current is proportional to the rate of change of that current. These two quantities are shown in Fig.4.4.

The applied voltage v is at all times equal and opposite to the back e.m.f. and this too is shown on Fig.4.4 so that it can be compared with the current which it is seen to lead by 90°.

The Reactance of an Inductor = $X_L = 2\pi fL = \omega L$

where f = frequency in Hz
L = inductance in henrys
$\omega = 2\pi F$.

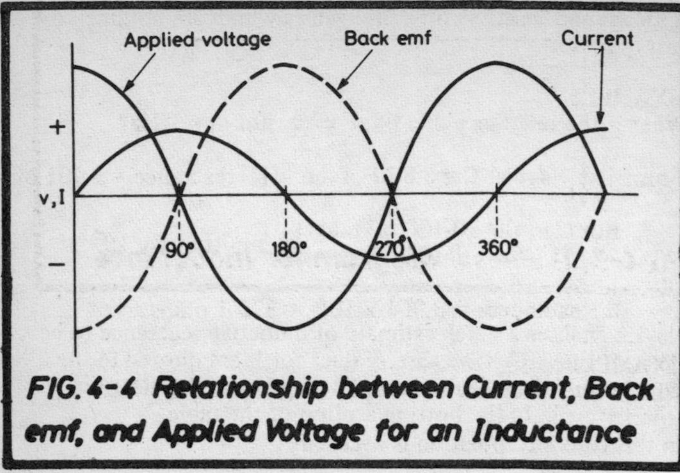

FIG. 4-4 Relationship between Current, Back emf, and Applied Voltage for an Inductance

X_L is directly proportional to frequency as shown in Fig.4.5(i) and conventionally is given a positive sign.

Fig.4.5(ii) shows the phasor diagram for a pure inductance, V_L representing the voltage applied and I_L the current.

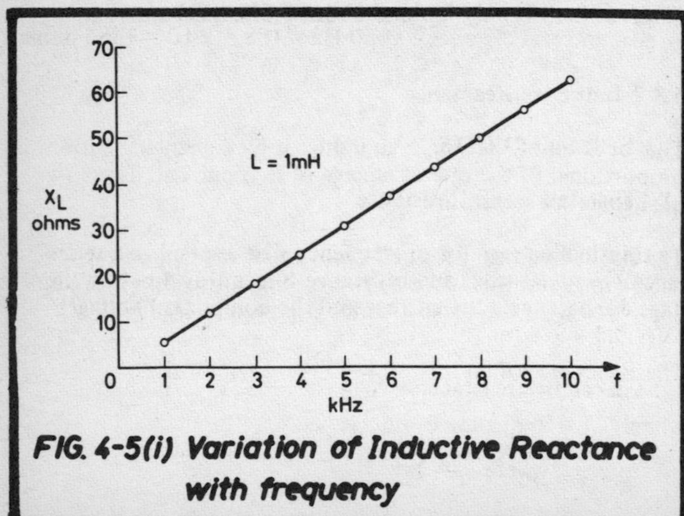

FIG. 4-5(i) Variation of Inductive Reactance with frequency

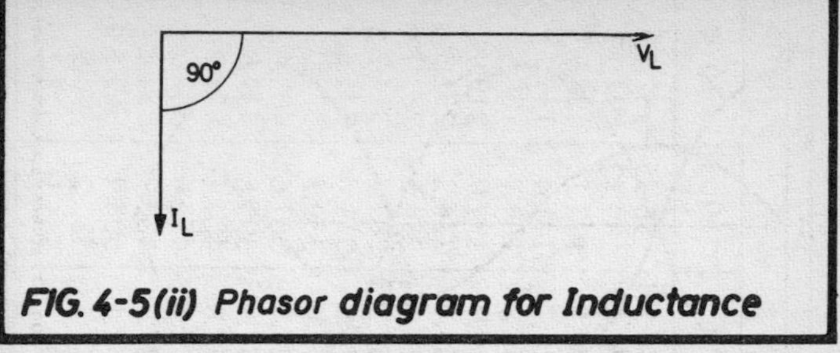

FIG. 4-5(ii) Phasor diagram for Inductance

Table 4.2 enables a quick estimate of inductive reactance to be made as in the capacitive case. Values for L are quoted in the range 1-10 mH with f in the range 1-10 kHz. Multiplication factors are given in the bottom section of the table.

EXAMPLE:
What is the reactance of a 6H choke at 50 Hz?

From Table 4.2 for 6 mH at 5 kHz $X_L = 188.5$

For L in the 1-10 H range,
f in the 10-100 Hz range multiply by 10.

Then reactance of 6H choke at 50 Hz = 188.5 x 10 = **1885 ohms**

4.4 SERIES CIRCUITS

4.4.1 Capacitance and Resistance

Let a sinusoidal generator of frequency f be applied to a series circuit of capacitance C and resistance R as in Fig.4.6. V is the voltage applied and I the current and the component voltages are V_C and V_R.

Let the capacitive reactance = X_C

$$= \frac{1}{\omega C}, \text{ where } \omega = 2\pi f$$

Inductance L (mH)	FREQUENCY f (kHz)									
	1	2	3	4	5	6	7	8	9	10
1.0	6.283	12.57	18.85	25.13	31.42	37.70	43.98	50.27	56.55	62.83
1.5	9.425	18.85	28.27	37.70	47.12	56.55	65.97	75.40	84.82	94.25
2.0	12.57	25.13	37.70	50.27	62.83	75.40	87.96	100.5	113.1	125.7
2.5	15.71	31.42	47.12	62.83	78.54	94.25	110.0	125.7	141.4	157.1
3.0	18.85	37.70	56.55	75.40	94.25	113.1	132.0	150.8	169.7	188.5
3.5	21.99	43.98	65.97	87.96	110.0	132.0	153.9	175.9	197.9	219.9
4.0	25.13	50.27	75.40	100.5	125.7	150.8	175.9	201.1	226.2	251.3
4.5	28.27	56.55	84.82	113.1	141.4	169.7	197.9	226.2	254.5	282.7
5.0	31.42	62.83	94.25	125.7	157.1	188.5	219.9	251.3	282.7	314.2
5.5	34.56	69.11	103.7	138.2	172.8	207.3	241.9	276.5	311.0	345.6
6.0	37.70	75.40	113.1	150.8	188.5	226.2	263.9	301.6	339.3	377.0
6.5	40.84	81.68	122.5	163.4	204.2	245.0	285.9	326.7	367.6	408.4
7.0	43.98	87.96	131.9	175.9	219.9	263.9	307.9	351.9	395.8	439.8
7.5	47.12	94.25	141.4	188.5	235.6	282.7	329.9	377.0	424.1	471.2
8.0	50.27	100.5	150.8	201.1	251.3	301.6	351.9	402.1	452.4	502.7
8.5	53.41	106.8	160.2	213.6	267.0	320.4	373.9	427.3	480.7	534.1
9.0	56.55	113.1	169.6	226.2	282.7	339.3	395.8	452.4	508.9	565.5
9.5	59.69	119.4	179.1	238.8	298.5	358.1	417.8	477.5	537.2	596.9
10.0	62.83	125.7	188.5	251.3	314.2	377.0	439.8	502.7	565.5	628.3

Multiply Table Figures for Ranges of f and L as shown:

f → / L ↓	10-100 Hz	100 Hz -1 kHz	1-10 kHz	10-100 kHz	100 kHz -1 MHz	1-10 MHz	10-100 MHz	100 MHz -1 GHz
1-10 μH	$\div 10^5$	$\div 10^4$	$\div 10^3$	$\div 10^2$	$\div 10$	NIL	x 10	x 10^2
10-100 μH	$\div 10^4$	$\div 10^3$	$\div 10^2$	$\div 10$	NIL	x 10	x 10^2	x 10^3
100 μH-1 mH	$\div 10^3$	$\div 10^2$	$\div 10$	NIL	x 10	x 10^2	x 10^3	x 10^4
1-10 mH	$\div 10^2$	$\div 10$	NIL	x 10	x 10^2	x 10^3	x 10^4	x 10^5
10-100 mH	$\div 10$	NIL	x 10	x 10^2	x 10^3	x 10^4	x 10^5	x 10^6
100 mH-1 H	NIL	x 10	x 10^2	x 10^3	x 10^4	x 10^5	x 10^6	x 10^7
1-10 H	x 10	x 10^2	x 10^3	x 10^4	x 10^5	x 10^6	x 10^7	x 10^8
10-100 H	x 10^2	x 10^3	x 10^4	x 10^5	x 10^6	x 10^7	x 10^8	x 10^9

NIL – indicates no correction

Table 4.2 **Inductive Reactance Values (ohms)**

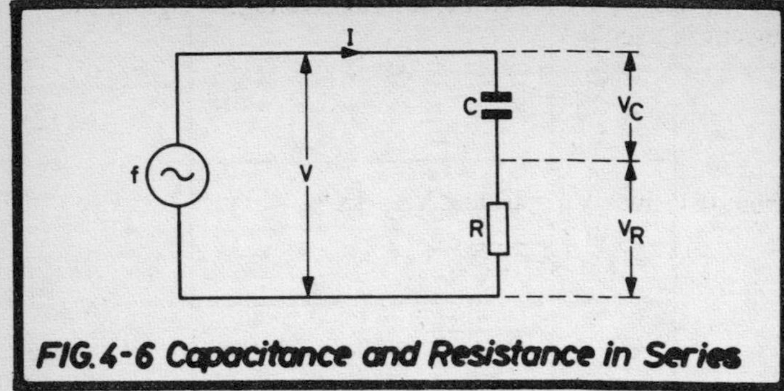

FIG.4-6 Capacitance and Resistance in Series

The phasor diagram is set up using as a reference, a line representing I because this is common to both C and R – the line is OI in Fig.4.7.

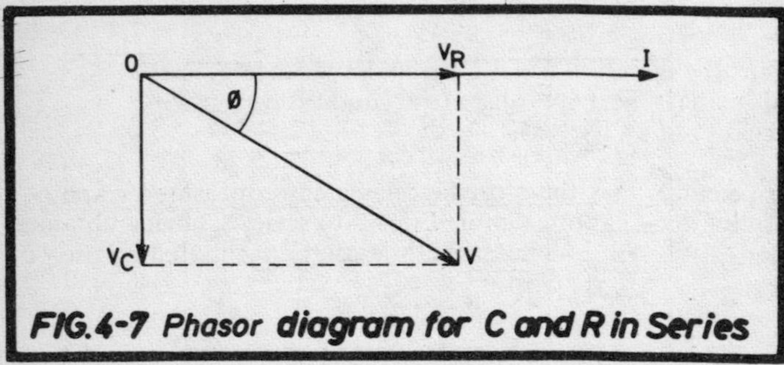

FIG.4-7 Phasor diagram for C and R in Series

V_R is in phase with I and is marked on OI according to its value $V_R = IR$.

$V_C = IX_C$ is also drawn to scale at right angles to OI, lagging by 90°.

Then the resultant of V_C and V_R is the phasor OV and from Pythagoras' Theorem

$$(OV)^2 = (OV_R)^2 + (OV_C)^2 \text{ equivalent to } V^2 = V_R^2 + V_C^2$$

in the circuit

$$\text{Hence } V = \sqrt{V_R^2 + V_C^2} \qquad (1)$$

and the angle ϕ by which the resultant voltage V lags on the current is given by

$$\tan \phi = \frac{V_C}{V_R} \qquad (2)$$

From (1), since $V_R = IR$ and $V_C = IX_C$

$$V = I\sqrt{R^2 + X_C^2}$$

$$\therefore \frac{V}{I} = Z = \sqrt{R^2 + X_C^2}$$ where Z is the ratio of voltage

to current for the series circuit and is known as the impedance.

From (2) $\tan \phi = \dfrac{X_C}{R}$

Thus the circuit of Fig.4.6 has an impedance of $\sqrt{R^2 + X_C^2}$ with a negative phase angle (the circuit is capacitive) equal to $\tan^{-1} (X_C/R)$.

Impedances may therefore be calculated or measured graphically by drawing for example Fig.4.7 to scale, X_C being obtained from Table 4.1. The latter method can be facilitated by use of a special chart, see Sect.4.6.

4.4.2 Inductance and Resistance

Fig.4.8 shows a resistance and inductance in series with both voltages and currents as marked.

Let the inductive reactance = X_L (= ωL, where $\omega = 2\pi f$).

As in the capacitive case (Sect.4.4.1), the current is common to both components and OI is drawn as the reference phasor as in Fig.4.9.

V_R is in phase with I and is marked on OI according to its value, $V_R = I_R$.

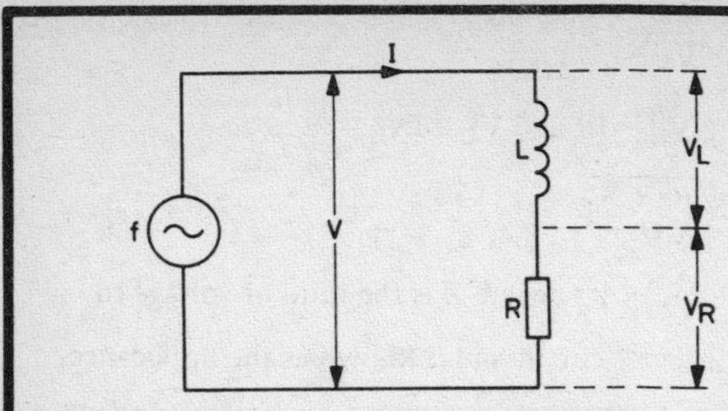

FIG. 4-8 Inductance and Resistance in Series

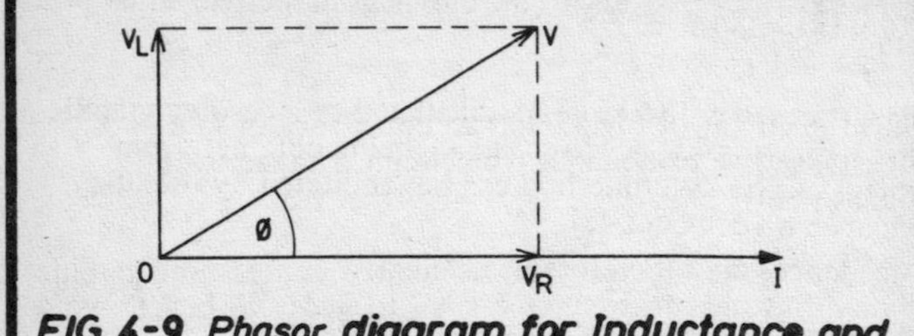

FIG. 4-9 Phasor diagram for Inductance and Resistance in Series

$V_L = IX_L$ is also drawn to scale at right angles to OI, leading by 90°.

Then the resultant of V_L and V_R is the phasor OV and from Pythagoras' Theorem

$$(OV)^2 = (OV_R)^2 + (OV_L)^2 \text{ equivalent to } V^2 = V_R^2 + V_L^2$$

in the circuit

$$\text{Hence } V = \sqrt{V_R^2 + V_L^2} \qquad (1)$$

and the angle ϕ by which the resultant voltage V leads the current is given by

$$\tan \phi = \frac{V_L}{V_R} \qquad (2)$$

From (1) since $V_R = IR$ and $V_L = IX_L$

$$V = I\sqrt{R^2 + X_L^2}$$

$\therefore \frac{V}{I} = Z = \sqrt{R^2 + X_L^2}$ where Z is the ratio of voltage to current for the series circuit and is known as the impedance.

From (2) $\tan \phi = \frac{X_L}{R}$

Thus the circuit of Fig.4.8 has an impedance of $\sqrt{R^2 + X_L^2}$ with a positive phase angle (the circuit is inductive) equal to $\tan^{-1} (X_L/R)$.

Impedances may therefore be calculated or measured graphically by drawing Fig.4.9 to scale, X_L being obtained from Table 4.2. The latter method can be facilitated by use of a special chart, see Sect.4.6.

4.4.3 Capacitance, Inductance and Resistance

The reasoning here follows directly from the two preceding sections (4.4.1 and 4.4.2).

Because the current leads the voltage in a purely capacitive circuit and lags in a purely inductive one, in both cases by 90°, it is clear from Fig.4.10 which shows a phasor diagram of a hypothetical circuit having a combination of both, that they are in direct opposition and therefore in the diagram subtractive, the net arithmetical reactance being positive or negative according to which of the component reactances is the greater.

In Fig.4.10 V_L and V_C are first.drawn as in the preceding sections and the equivalent phasor diagram follows where V_L has been subtracted from V_C, leaving the circuit with a net capacitive reactance and a small resultant voltage V lagging on the current and given by

$$V = I(X_C - X_L)$$

But this is not a practical circuit therefore R must be introduced (at least to account for the winding resistance of the inductor) which gives a further voltage component as before in phase with the current, the resultant phasor diagram being re-drawn in Fig.4.11.

Thus, as reasoned in the preceding sections:

$$Z = \sqrt{R^2 + (X_L - X_C)^2} \quad \text{and} \quad \phi = \tan^{-1} \frac{(X_L - X_C)}{R}$$

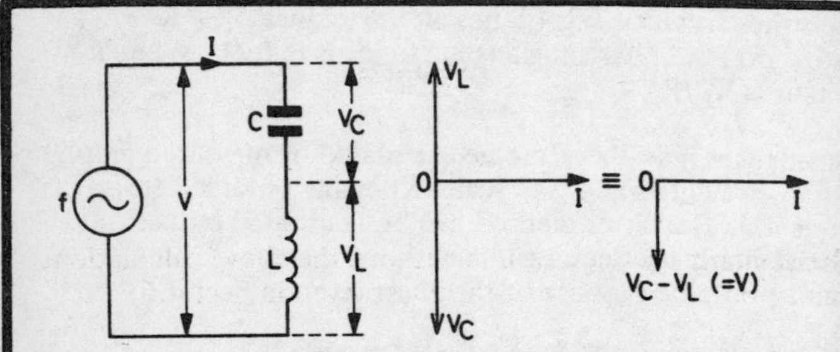

FIG. 4-10 Phasor diagrams for Capacitance and Inductance in Series

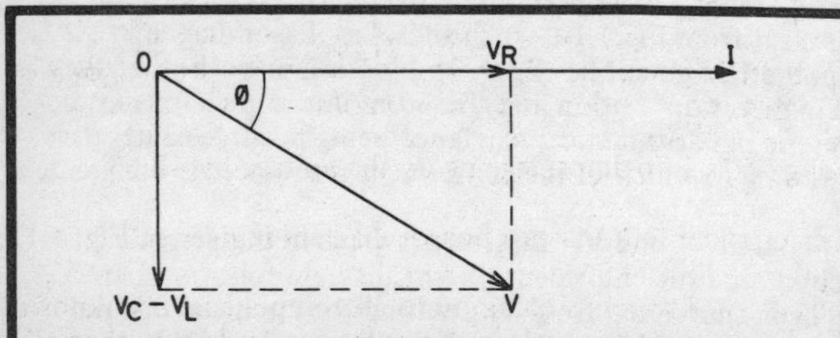

FIG. 4-11 Resistance added to FIG. 4-10

When X_C happens to be greater than X_L (as in Fig.4.11), Z is positive because the square of $(X_L - X_C)$ is positive, but ϕ is negative showing that the voltage lags on the current.

EXAMPLE:
What is the impedance of a series combination of resistor, 100 ohms, capacitor, 20 μF and inductor, 0.8 H at 50 Hz? What current flows if a voltage of 240 is applied at this frequency?

$$R = 100 \text{ ohms}, \quad C = 20\,\mu F, \quad L = 0.8\,H$$

From Table 4.1 X_C at 50 Hz = 159.2 ohms
" Table 4.2 X_L at 50 Hz = 251.3 ohms

Then $(X_L - X_C)$ = 92.1 ohms

$$Z = \sqrt{100^2 + 92.1^2} = \mathbf{136\ ohms}$$

$$\phi = \tan^{-1}\frac{92.1}{100} = \mathbf{42.65°}$$

(Reasonably accurate results without the above calculations can be obtained by use of the chart given in Sect.4.6).

Current for 240V applied:

$$I = \frac{V}{Z} = \frac{240\,\angle 0°}{136\,\angle 42.65°} = \mathbf{1.76\,\angle{-42.65°}}$$

(Note Multiplication of vectors: $r_1\angle\phi_1 \times r_2\angle\phi_2 = r_1 r_2\angle\phi_1 + \phi_2$

Division " " $\dfrac{r_1\angle\phi_1}{r_2\angle\phi_2} = \dfrac{r_1}{r_2}\angle\phi_1 - \phi_2$

ie current is 1.76 amps, lagging on the voltage by 42.65°.

The complete solution as a phasor diagram is given in Fig. 4.12.

The current I which is common to all components is unknown. Thus any convenient scale is chosen eg for I = 1 amp, then all voltages take on the same numerical values as the resistances or

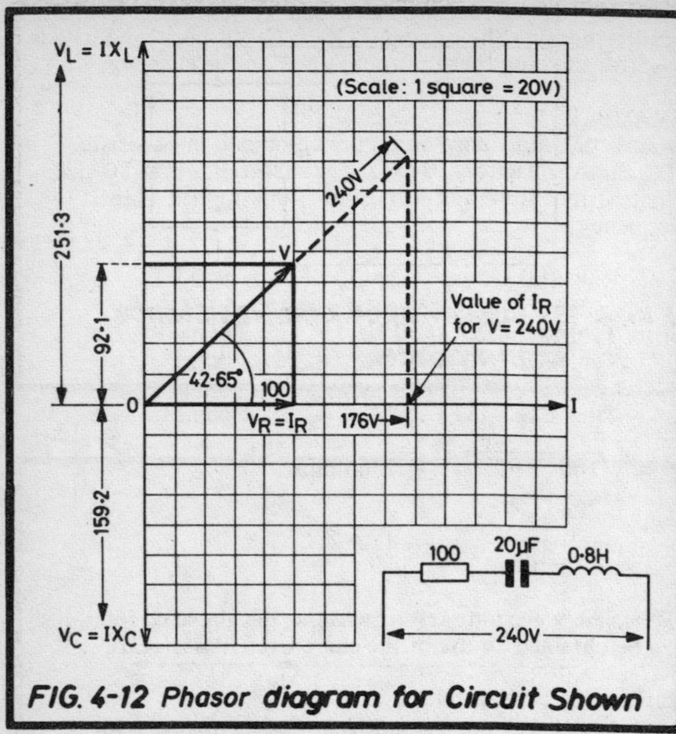

FIG. 4-12 Phasor diagram for Circuit Shown

reactances across which they are developed. By drawing these, the phasor V is produced at the correct angle to O1 but for a current of 1 amp. However, only the angle between V and I is necessary to determine one if the other is known, thus 240V is scaled onto the phasor for V from which, as shown, the value of IR is determined, hence of I by dividing by R.

4.5 PARALLEL CIRCUITS

4.5.1 Capacitance and Resistance

A resistor in parallel with a capacitor is shown in Fig.4.14 with appropriate currents and voltage marked.

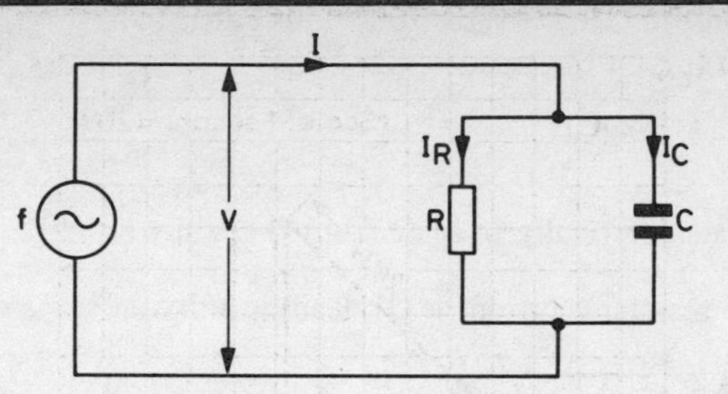

FIG. 4-13 Capacitance and Resistance in Parallel

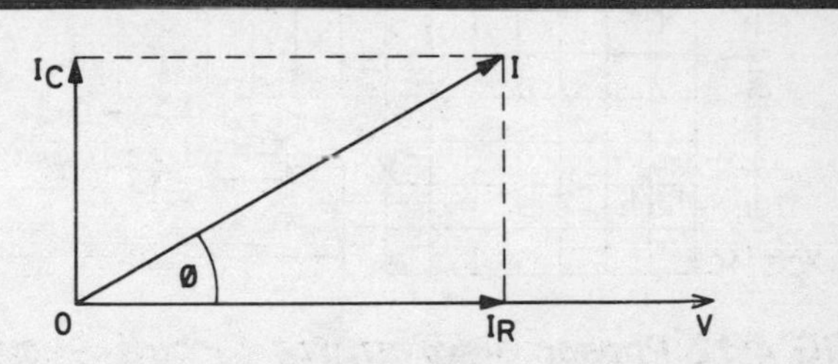

FIG. 4-14 Phasor diagram for Capacitance and Resistance in Parallel

Let $X_C = \frac{1}{\omega C}$ be the reactance of the capacitor at frequency f.

The quantity common to both C and R is the voltage V and the phasor diagram can be drawn with OV as the reference as in Fig.4.14.

The current in R is in phase with V and is therefore marked along OV

$$\text{as} \quad I_R = \frac{V}{R}$$

The current in C leads V by 90° and is equal to

$$\frac{V}{X_C} \ (= V\omega C)$$

this is drawn vertically to scale from O as shown.

Then the resultant current is OI, leading V by an angle ϕ.

From Fig.4.14

$$I^2 = I_R^2 + I_C^2$$

$$\therefore \ I = \sqrt{\left(\frac{V}{R}\right)^2 + \left(\frac{V}{X_C}\right)^2}$$

$$\therefore \ \frac{I}{V} = \frac{1}{Z} = \sqrt{\frac{1}{R^2} + \frac{1}{X_C^2}}$$

$$\text{from which} \quad Z = \frac{R.X_C}{\sqrt{R^2 + X_C^2}} \qquad (1)$$

$$\text{and} \quad \tan\phi = \frac{I_C}{I_R} = \frac{\frac{V}{X_C}}{\frac{V}{R}} = \frac{R}{X_C} \quad \text{ie } \phi = \tan^{-1}\frac{R}{X_C} \qquad (2)$$

X_C can be calculated from $\frac{1}{2\pi fC}$ or obtained directly from Table 4.1 and hence Z and ϕ calculated.

4.5.2 Inductance and Resistance

An inductor in parallel with a resistor is shown in Fig.4.15 with appropriate currents and voltage marked.

Let $X_L (= \omega L)$ be the reactance of the inductor at frequency f.

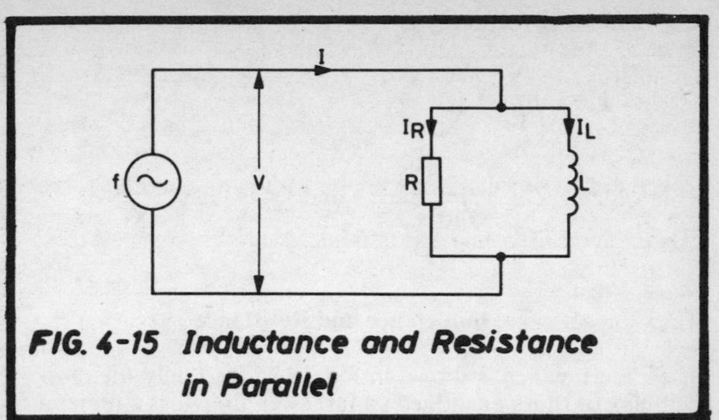

FIG. 4-15 Inductance and Resistance in Parallel

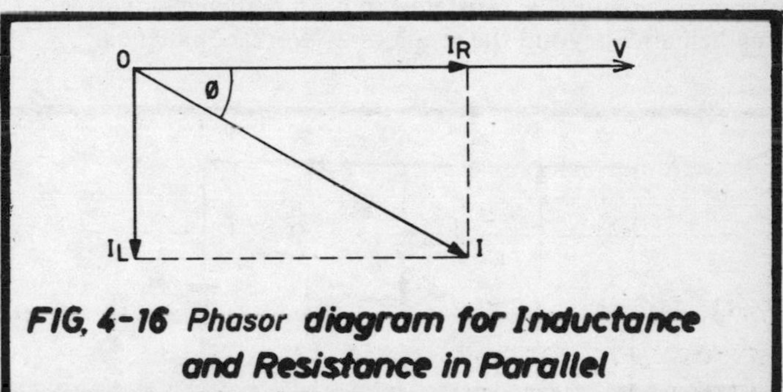

FIG. 4-16 Phasor diagram for Inductance and Resistance in Parallel

As with capacitance in the previous section, the quantity common to both components is the voltage V, and the phasor diagram is set up on OV as the reference as in Fig.4.16.

$$I_R = \frac{V}{R} \text{ is in phase with V,}$$

$$I_L = \frac{V}{X_L} \text{ lags on V by } 90°.$$

The resultant current is represented by OI and

$$I^2 = I_R^2 + I_L^2$$

$$\therefore\ I = \sqrt{\frac{V}{R^2} + \frac{V}{X_L^2}}$$

from which $Z = \dfrac{RX_L}{\sqrt{R^2 + X_L^2}}$ and $\phi = \tan^{-1} \dfrac{R}{X_L}$

4.5.3 Capacitance, Inductance and Resistance

The circuit, which is shown in Fig.4.17 is certainly the most complex of those examined so far, even though as a representation of a practical parallel circuit, no resistance has been allowed for in the capacitive branch. In fact full analysis of the complete circuit (i.e. resistance in both branches) involves mathematics beyond the scope of a "workshop manual".

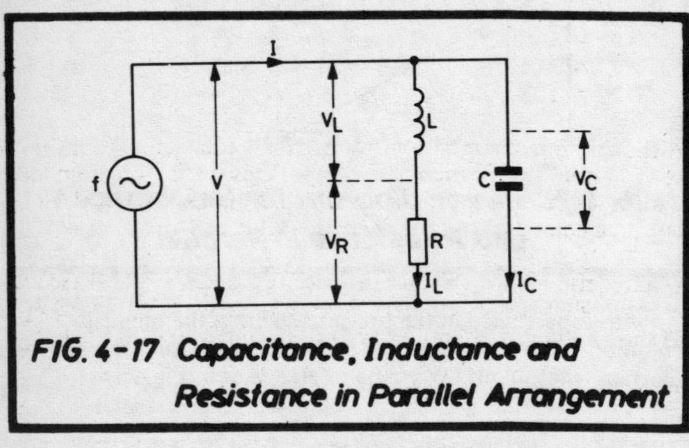

FIG. 4-17 Capacitance, Inductance and Resistance in Parallel Arrangement

However, the circuit as shown is an extremely important one because of its inestimable value when resonant, (see Sect. 4.7). It is not proposed to examine the circuit in great depth in this section as many of its features are brought out in the later one. Sufficient here to quote the impedance formulae and to go through the steps of the much more practical phasor diagram.

$$|Z| = \frac{1}{\sqrt{\left(\frac{R}{R^2 + \omega^2 L^2}\right)^2 + \left(\frac{\omega L}{R^2 + \omega^2 L^2} - \omega C\right)^2}}$$

$$\phi = \tan^{-1} \frac{\left(\frac{\omega L}{R^2 + \omega^2 L^2} - \omega C\right)}{\left(\frac{R}{R^2 + \omega^2 L^2}\right)}$$

Phasor Diagram

A practical example will show the simplicity and clarity of the method compared with calculation.

Suppose the values are as follows, with reactances from Tables 4.1 and 4.2

f = 200 kHz
C = 100 pF, ∴ X_C = 7960 ohms
L = 3 mH, ∴ X_L = 3770 ohms
R = 1000 ohms

Although V is common to both parallel paths, its relationship to I_L is not known, therefore it is perhaps better to commence with I_L as the reference phasor which is marked in Fig. 4.18 as OI_L, for convenience I_L is given the value of 1 mA.

I_L gives rise to two voltages in series, V_R and V_L and these are accordingly constructed to scale on OI_L giving OV_R and OV_L as shown, their resultant being OV, leading I_L by the angle ϕ_1.

By measurement V = 3.9 volts and $\phi_1 = 75°$.

At this stage, having obtained the relationship between V and I_L, V becomes a second reference phasor and the relationship between I_L and I_C has to be determined to obtain the total current I.

Since $I_C = \frac{V}{X_C}$, this is calculated to be 0.49 mA which is

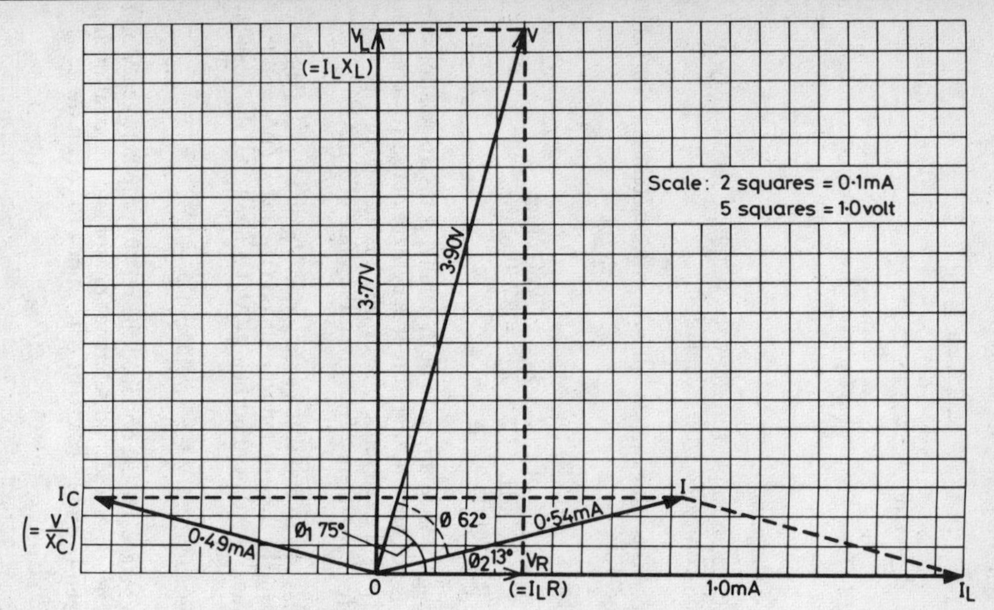

FIG. 4-18 Phasor diagram for Parallel circuit of FIG. 4-17

marked out as the phasor OI_C leading V by 90°. Completion of the parallelogram on OI_L and OI_C enables the phasor OI to be drawn which measures 0.54 mA lagging on V by 62°.

So, from the phasor diagram:

$$Z = \frac{V}{I} = \frac{3.9 \angle 75^\circ}{(0.54 \times 10^{-3}) \angle 13^\circ} = 7220 \angle 62^\circ \text{ ohms.}$$

Summing up, the phasor diagram gives a complete picture of all the relative circuit conditions at once – for I_L = 1 mA. For any other values, the scale is altered or the values multiplied or divided accordingly e.g. for an applied voltage V = 0.1v, all phasor values are divided by 39. The diagram is equally useful for a study of the effects of changes which may be made.

4.6 SIMPLIFIED IMPEDANCE EVALUATION

The impedance triangle can be readily solved by use of the special chart of Fig.4.19. Although slightly more approximate than the careful drawing, the accuracy is usually sufficient and furthermore it enables an especially quick answer to be obtained. It is also of use as a rapid check on calculations of impedance to ensure that decimal points, noughts etc. have not be misplaced.

The intersection of lines drawn from the R and X axes at the appropriate values marks the end of the vector for Z, drawn from O. The magnitude of Z is scaled by the circle quadrants which have their values marked on both the R and the X axes. Some interpolation between the nearest circle quadrants may be necessary but note that the distance between adjacent ones along any impedance vector is on the same scale as R or X and therefore a small piece of card or paper may be scaled for easy sub-division. The angle of the impedance is read off the circle quadrant so marked.

To take a simple, known example, let R = 30 ohms, X = 40 ohms – their two vectors intersect on the circle quadrant marked 50 ohms (as would be expected from a 3, 4, 5 triangle).

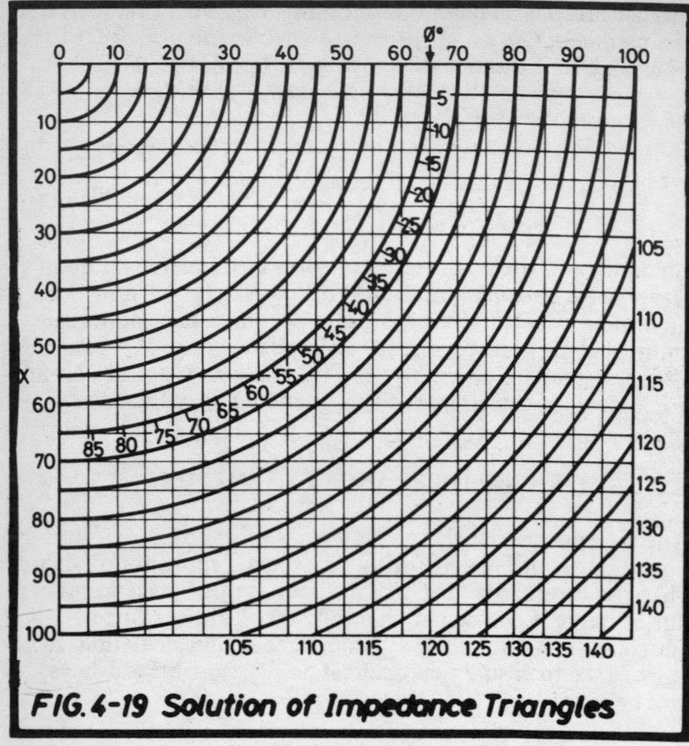

FIG. 4-19 Solution of Impedance Triangles

All values may be multiplied or divided together by multiples of 10 as required and this caters for all combinations of R and X when the ratio between them is less than 10:1. If the ratio happens to be greater than 10:1, calculation is probably unnecessary because the value of Z will be found to lie within 0.5% of R or X whichever is the greater and the angle is very small, ie the circuit is almost wholly resistive or reactive.

EXAMPLE:
A resistor of 200 ohms is connected in series with a capacitor of 0.1 μF. What is the impedance of the combination at 5 kHz?

From Table 4.1 $X_C = 318.3$ ohms

" Fig.4.19 at $R = 20$, $X = 31.8$, $Z = 37.5$
$\therefore$ for $R = 200$, $X = 318$, $Z = 375$.

Also $\phi = 58°$. Since the circuit is capacitive, the angle is negative,

$$\therefore \text{Impedance of combination} = 375 \angle{-58°}$$

4.7 RESONANT CIRCUITS

4.7.1 Resonant Frequency

A second glance at Fig.4.11 shows that an interesting condition occurs in the C,L,R circuit when $V_C = V_L$ and hence $V_C - V_L = 0$, there is then no phasor at 90° lag or lead on the the reference phasor. The applied voltage and current are in phase and the impedance is at its lowest and equal to the resistance R. This is a series resonant circuit and resonance occurs when $X_L = X_C$ i.e.

$$\omega L = \frac{1}{\omega C}, \quad \omega = \frac{1}{\sqrt{LC}}$$

$$\therefore f_r = \frac{1}{2\pi\sqrt{LC}}$$

where f_r is the frequency of resonance.

Because X_L increases with frequency, and X_C falls, then at frequencies above f_r the circuit has a net inductive reactance (positive) conversely at frequencies below f_r there is a net capacitive reactance (negative).

The parallel circuit is more complicated, it has not the advantage of the series circuit in that V_L and V_C are directly additive. If the condition for reasonance is defined as that in which the current is in phase with the applied voltage, i.e. the circuit exhibits no net reactance, it will be found that X_L and X_C are not quite equal.

By use of operator j (complex algebra) the resonant frequency f_r can be most easily established, see Fig.4.17

$$I = I_L + I_C = \frac{V}{R + j\omega L} + j V\omega C$$

which, by rearrangement and equating the j terms to zero (i.e. no net reactance) gives

$$f_r = \frac{1}{2\pi}\sqrt{\frac{1}{LC} - \frac{R^2}{L^2}}$$

It may help in understanding the complexities of the voltages and current conditions if the exercise of Sect.4.5.3 is repeated for the resonant frequency, see Fig.4.20. The component values are unchanged but the frequency has been altered to the resonant one, 285.7 kHz.

Again, choosing 1 mA for I_L, the reference phasor is determined

V_L (= IX_L = 5.385 V) is drawn, leading I_L by 90°
V_R (= IR = 1.0 V) is marked along OI_L, being in phase.

Then OV is the resultant voltage phasor, with measured value 5.47 V.

On OV as a second reference, $I_C = \frac{V}{X_C} = 0.982$ mA is drawn, leading by 90°

Completion of the parallelogram O,I_C,I,I_L, shows that I, the resultant of I_L and I_C falls on the phasor OV, i.e. the applied voltage and the current it produces are in phase.

It will be further seen that if, for example, a coil were available of known values of L and R, and the value of C were required for tuning to a given frequency, this could be determined from a phasor diagram of the type of Fig. 4.20. The unknown phasor would be the length of OI_C (at right angles to the phasor V) which would be obtained by completion of the O,I_C,I,I_L, parallelogram. From the length of OI_C, converted to mA, the value of C may be determined:

$$C = \frac{I_C}{\omega V} \times 10^{-3} \text{ farads} = \frac{I_C}{\omega V} \times 10^{9} \text{ pF}$$

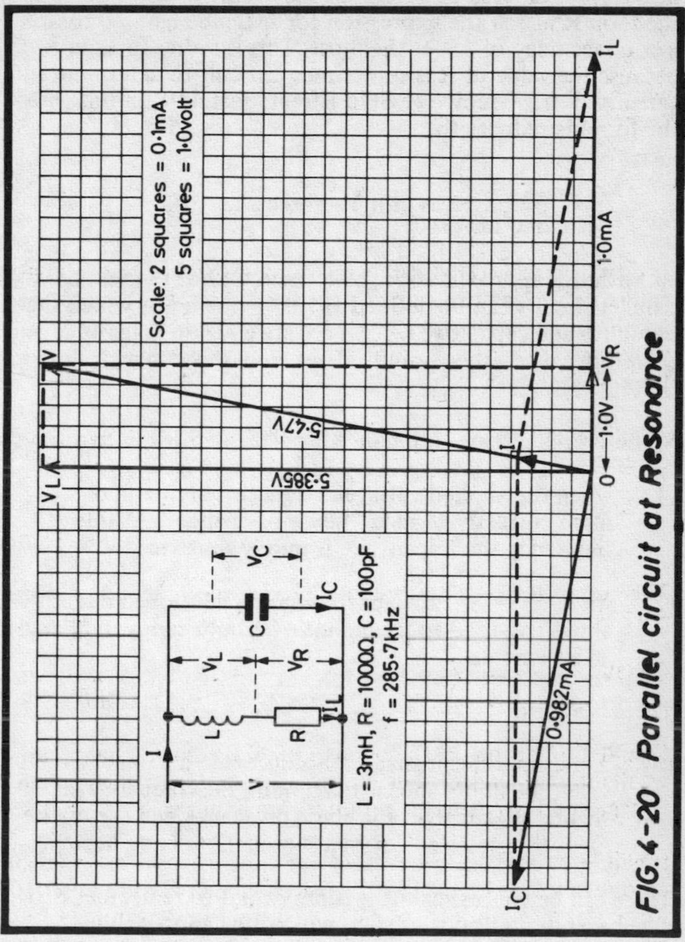

FIG. 4-20 Parallel circuit at Resonance

Approximations

In the above example, a high value for R has been used simply to aid demonstration by vector diagram. In practice the resistance of the inductor would more likely be in the 30-50 ohm range at the resonant frequency. It can be shown that provided that R is not greater than about 0.1 of ωL (and in the practical case of 50 ohms, this is only about 0.01 of ωL), the term R^2/L^2 in the expression for f_r can be ignored for a loss of accuracy of 1% at the most. This is rather fortunate because the value of R is sometimes difficult to determine and varies with frequency (see Skin Effect, Sect.4.9). In this case the formula reduces to

$$f_r = \frac{1}{2\pi\sqrt{LC}} \quad \text{as in the series case,}$$

so, with this approximation, resonance for both series and parallel circuits can be defined as the frequency at which the inductive and capacitive reactances are equal in magnitude and therefore cancel (the formula arises from this concept, Sect.4.7.1).

So far in this section, resonance has been defined in two ways:

(1) when the applied voltage and its current are in phase, more correctly termed "phase resonance". At this frequency the impedance is purely resistive

(2) when $X_L = X_C$, the approximation developed directly above and used to generalize over both series and parallel arrangements.

There is a third:

(3) Amplitude Resonance – this occurs in a parallel circuit when the impedance is maximum, the frequency is slightly higher than for phase resonance and the mathematical expression for this is even more complicated.

These can be confusing but it is important to remember that whichever definition is used, f_r will not be appreciably different, therefore in practical work, very little is lost if the differences are ignored.

Table 4.3 gives values of f_r for a wide range of values of L and C when these commence with a single integer – if the actual values lie between the table values, the table is still useful as a guide or as a check on calculations.

EXAMPLE:
What is the approximate resonant frequency of the circuit shown in Fig.4.17 if C = 100 pF, L = 3 mH and R is small?

From Table 4.3 at C = 1, L = 3

A figure = 91.888 B figure = 290.58

From lower table for L within 1-10 mH range and C within 100-1000 pF range, multiple B figure by 10^3.

$\therefore$ Resonant frequency = **290.58 kHz**

Note that the range for C could have been chosen as 10-100 pf, in which case the figures at the bottom of the first table would be used, these give

A figure 29.058 B figure 91.888

Then, from lower table for L within 1-10 mH range and C within 10-100 pF range, multiply A figure by 10^4, giving the same answer.

4.7.2 Impedance at Resonance

Series Circuit

At resonance a series circuit has minimum impedance, which is, in fact resistive and equal to the value of the resistance. From Sect.4.4.3

$$Z = \sqrt{R^2 + (X_L - X_C)^2} \quad \phi = \tan^{-1} \frac{(X_L - X_C)}{R}$$

but since at reasonance $X_L = X_C$, $(X_L - X_C) = 0$

Then Impedance at Resonance: $Z_r = R \quad \phi_r = 0°$

L → / C ↓	1		2		3		4		5	
	A	B	A	B	A	B	A	B	A	B
1	159.15	503.29	112.54	355.88	91.888	290.58	79.578	251.65	71.175	225.08
2	112.54	355.88	79.577	251.64	64.974	205.47	56.270	177.94	50.329	159.15
3	91.888	290:58	64.974	205.47	53.052	167.76	45.944	145.29	41.093	129.95
4	79.578	251.64	56.270	177.94	45.944	145.29	39.789	125.82	35.588	112.54
5	71.175	225.07	50.329	159.15	41.093	129.95	35.588	112.54	31.831	100.66
6	64.974	205.47	45.944	145.29	37.514	118.63	32.487	102.73	29.058	91.888
7	60.155	190.23	42.535	134.51	34.730	109.83	30.077	95.113	26.902	85.071
8	56.270	177.94	39.789	125.82	32.487	102.73	28.135	88.971	25.165	79.578
9	53.052	167.76	37.514	118.63	30.630	96.859	26.526	83.881	23.725	75.026
10	50.329	159.15	35.588	112.54	29.058	91.888	25.165	79.578	22.508	71.175

L → / C ↓	6		7		8		9		10	
	A	B	A	B	A	B	A	B	A	B
1	64.974	205.47	60.155	190.23	56.270	177.94	53.052	167.76	50.329	159.15
2	45.944	145.29	42.535	134.51	39.789	125.82	37.514	118.63	35.588	112.54
3	37.514	118.63	34.730	109.83	32.487	102.73	30.630	96.859	29.058	91.888
4	32.487	102.73	30.077	95.113	28.135	88.971	26.526	83.881	25.165	79.578
5	29.058	91.888	26.902	85.071	25.165	79.578	23.725	75.026	22.508	71.175
6	26.526	83.881	24.558	77.660	22.972	72.643	21.658	68.489	20.547	64.974
7	24.558	77.660	22.736	71.898	21.268	67.255	20.052	63.409	19.023	60.155
8	22.972	72.643	21.268	67.255	19.894	62.911	18.757	59.313	17.794	56.270
9	21.658	68.489	20.052	63.409	18.757	59.313	17.684	55.921	16.776	53.052
10	20.547	64.974	19.023	60.155	17.794	56.270	16.776	53.052	15.915	50.329

MULTIPLY COLUMN A or B FIGURES AS SHOWN:

L → / C ↓	1-10 μH		10-100 μH		100-1000 μH		1-10 mH		10-100 mH		100-1000 mH		1-10 H	
	A	B	A	B	A	B	A	B	A	B	A	B	A	B
1-10 pF	10^6			10^5	10^5			10^4	10^4			10^3	10^3	
10-100 pF		10^5	10^5			10^4	10^4			10^3	10^3			10^2
100-1000 pF	10^5			10^4	10^4			10^3	10^3			10^2	10^2	
0.001-0.01 μF		10^4	10^4			10^3	10^3			10^2	10^2			10
0.01-0.1 μF	10^4			10^3	10^3			10^2	10^2			10	10	
0.1-1.0 μF		10^3	10^3			10^2	10^2			10	10			NIL
1-10 μF	10^3			10^2	10^2			10	10			NIL	NIL	

$$f_r = \frac{1}{2\pi\sqrt{LC}} \text{ Hz}$$

where f_r = resonant frequency
L = Inductance in Henrys
C = Capacitance in Farads.

Table 4.3 Resonant Frequencies (Hz) for Range of Values of L and C

Parallel Circuit

Consider the circuit of Fig.4.17 (i.e. an inductance with its series resistance in parallel with a capacitor, assumed to have no loss).

As shown in the previous section:

$$I = \frac{V}{R + j\omega L} + jV\omega C$$

At resonance it is only necessary to solve for the in-phase component (since I has no reactive component).

$$\text{Then} \quad I = \frac{VR}{R^2 + \omega_r^2 L^2} \quad \text{where } \omega_r = 2\pi f_r$$

$$Z = \frac{V}{I} = \frac{R^2 + \omega_r^2 L^2}{R}$$

$$\text{But} \quad f_r = \frac{1}{2\pi}\sqrt{\frac{1}{LC} - \frac{R^2}{L^2}} \qquad \therefore \omega_r^2 = \frac{1}{LC} - \frac{R^2}{L^2}$$

$$\therefore \text{Impedance at Resonance } Z_r = \frac{L}{CR} \text{ ohms}$$

which is a pure resistance, i.e. $\phi_r = 0°$.

EXAMPLE:

(1) What is the impedance of the tuning circuit shown in Fig.4.21 at the resonant frequency of 1185.9 kHz?

Impedance at resonance, $Z_r = \dfrac{L}{CR}$

$$= \frac{90 \times 10^{-6}}{200 \times 10^{-12} \times 50}$$

$$= \mathbf{9{,}000 \text{ ohms}}$$

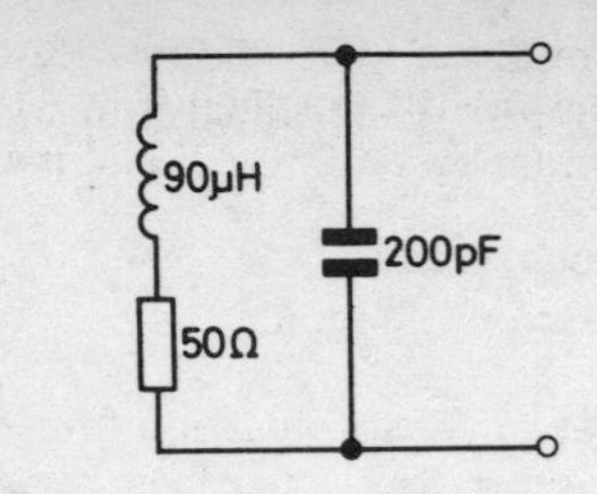

FIG. 4-21 Parallel Tuning circuit

(2) What is the resonant frequency and impedance at resonance if L, R and C are all connected in series?

$$f_r = \frac{1}{2\pi\sqrt{LC}} = \frac{1}{2\pi\sqrt{90 \times 10^{-6} \times 200 \times 10^{-12}}}$$

$$= \mathbf{1186.3\ kHz}$$

which again demonstrates that use of the series formula sacrifices little loss of accuracy.

Impedance at resonance Z_r = R = **50 ohms.**

This example demonstrates an important point with regard to impedances of series and parallel circuits, i.e. the former have a low impedance at resonance whereas the latter have a high impedance. In fact, as R tends to zero, the series impedance also tends to zero whereas the parallel one tends to infinity.

4.7.3 Q-Factor

This is a measure of the quality of a component or circuit when carrying an oscillating current. It is defined as

$$Q = \frac{2\pi \times (\text{maximum energy stored during one cycle})}{\text{Energy dissipated during one cycle}}$$

Q-Factor of Inductor

Let an inductor be represented by an inductance, L in series with a resistance, R. Suppose a sinusoidal (r.m.s.) current, I flows, then the maximum current = $\sqrt{2}I$ and maximum energy stored

$$= \tfrac{1}{2}L(\sqrt{2}I)^2 \text{ joules} = LI^2 \text{ joules}$$

and the energy lost in one cycle of duration t $\left(= \frac{1}{f}\right)$ secs.

$$= \frac{I^2R}{f} \text{ joules}$$

$$\text{Then } Q = \frac{2\pi LI^2 f}{I^2R} = \frac{\omega L}{R}$$

An idea of the variation of Q with frequency of an inductor is shown in Fig.4.22.

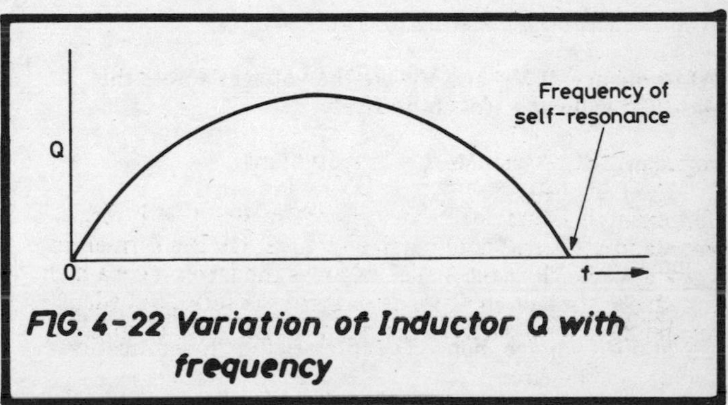

FIG. 4-22 Variation of Inductor Q with frequency

At low frequencies Q increases as ωL increases (i.e. directly proportional to frequency) but skin effect at higher frequencies causes R to increase approximately proportional to $\sqrt{f}$, thus Q varies as $f/\sqrt{f} = \sqrt{f}$. The curve follows this law until the shunting effect of the reactance of the capacitance between the coil winding turns (self-capacitance of the coil) increasingly reduces the effect of ωL until the latter becomes completely cancelled and Q falls to zero. At this point the inductor is self-resonant.

Q-Factor of Capacitor

Similar reasoning to that shown above gives

$$Q = \frac{1}{\omega CR}$$

Q-Factor of Series Resonant Circuit

$$Q = \frac{\omega_r L}{R} \qquad \text{where } f_r = \text{frequency of resonance}, \quad \omega_r = 2\pi f_r$$

$$= \frac{L}{\sqrt{LC}.R} \qquad \text{since } \omega_r = \frac{1}{\sqrt{LC}}$$

$$= \frac{1}{R}\sqrt{\frac{L}{C}}$$

Another important feature of Q follows:

At resonance, if V_L and V_C are the voltages across the inductor and capacitor respectively,

$$V_L = I\omega L = \frac{V\omega L}{R} = QV$$

$$\text{and } V_C = V_L = QV$$

i.e. the voltages across both inductor and capacitor are Q times the applied voltage, hence Q expresses the "magnification factor" of the circuit.

EXAMPLE:

If the tuning circuit of Fig.4.21 is arranged as a series circuit and a signal of 1 mV is applied from the aerial at the resonant frequency of 1186.3 kHz, what voltage is developed across the capacitor?

$$Q = \frac{1}{R}\sqrt{\frac{L}{C}}\left(\text{or } \frac{\omega_r L}{R}\right) = \frac{1}{50}\sqrt{\frac{90 \times 10^{-6}}{200 \times 10^{-12}}}$$

$$= 2\sqrt{45} = 13.42$$

$$\text{Then voltage across capacitor (or inductor)} = 1 \times 13.42 \text{ mV} = \mathbf{13.42\ mV}$$

i.e. an aerial signal of 1 mV has been magnified to over 13 mV, a much more useful signal to apply to subsequent stages of a receiver.

Generally, by making the not unreasonable approximation that the capacitor has no loss, at resonance, the Q-factor of the series circuit is equal to the Q-factor of the inductance.

Q-Factor of Parallel Resonant Circuit

Considering the parallel circuit with a no-loss capacitor (Fig.4.17), it can be shown that the maximum energy stored in the inductor is equal to the maximum stored in the capacitor. The Q-factor of an inductor has been shown to be equal to $\omega_r L/R$ at resonance, and it is also equal to the Q-factor of the parallel circuit.

Furthermore, provided that Q is not very low (less than about 10), the currents in the two branches are each approximately Q times the supply current, i.e. Q is a current "magnification factor".

Q-Factor of Quartz Crystals

If an electrical charge is applied in a certain manner to a quartz crystal, the piezo-electric effect results in a mechanical stress being produced. Thus if an alternating voltage is applied, the crystal will vibrate and if the frequency of the applied voltage is near to that at which mechanical resonance arises in the crystal, the amplitude of the vibration will be large. An equivalent electrical LCR circuit can be determined for any crystal such that the equivalent L and C resonate at the same frequency as the crystal. Thus Q-factor is appropriate to a quartz crystal as shown in the following typical example, see Fig.4.23

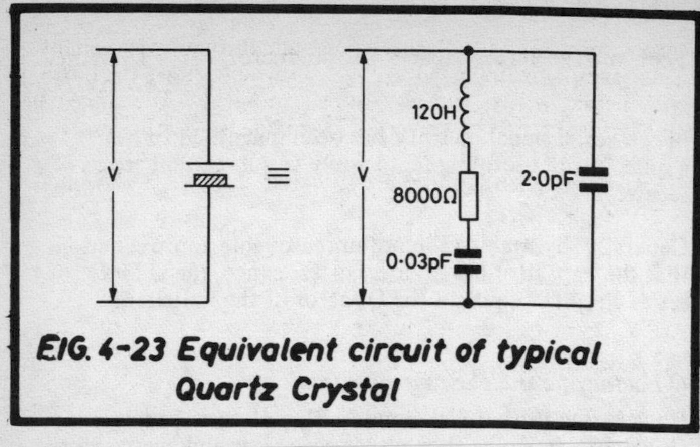

FIG. 4-23 Equivalent circuit of typical Quartz Crystal

$$f_r \text{ of series resonant circuit } = \frac{1}{2\pi\sqrt{LC}} = \frac{1}{2\pi\sqrt{120 \times .03 \times 10^{-12}}}$$

$$= 83.898 \text{ kHz}$$

$$Q = \frac{1}{R}\sqrt{\frac{L}{C}} = \frac{1}{8000}\sqrt{\frac{120}{.03 \times 10^{-12}}} = 7906$$

thus demonstrating the very high magnification factors (Q) obtainable, in fact, modern crystals are capable of Q-factors of up to 500,000. In addition, of course, the high stability of resonant frequency is a great asset.

4.7.4 Selectivity

Definition:

The degree to which a circuit which resonates at one frequency responds to other frequencies is known as selectivity. It is thus a measure of the rejection of frequencies other than the required one.

Series Circuits:

The response of a typical series circuit through its resonant frequency f_r is indicated in Fig.4.24 which shows the current I at constant voltage but at various frequencies. The change in phase angle between current and voltage is also shown.

The formulae for Z and ϕ are shown in Sect.4.4.3 and since $I = V/Z$

$$I = \frac{V}{\sqrt{R^2 + (X_L - X_C)^2}} \angle \tan^{-1}\left(\frac{X_L - X_C}{R}\right)$$

At resonance $X_L - X_C = 0$

thus the current at resonance, $I_r = \frac{V}{R} \angle 0^\circ$

At frequencies below f_r $X_C > X_L$ therefore current leads voltage
" " above f_r $X_L > X_C$ " current lags on "

At two particular frequencies f_b and f_a, below and above resonance

$$(X_L - X_C)^2 = R^2$$

$$\therefore I = \frac{V}{\sqrt{2R^2}} = \frac{\sqrt{2}V}{2R} = 0.707\,\frac{V}{R} = 0.707\,I_r$$

and the phase angle between I and V is $\tan^{-1} \pm 1 = \pm 45^\circ$

which easily determines two points other than I_r on the selectivity curve as shown in Fig.4.24.

The power at either f_b or f_a is given by

$$P = VI\cos\phi$$

and the circuit power ratio at f_b or f_a compared with f_r is

$$\frac{VI\cos\phi}{VI_r\cos\phi_r}$$ where $I\cos\phi$ relates to f_b or f_a

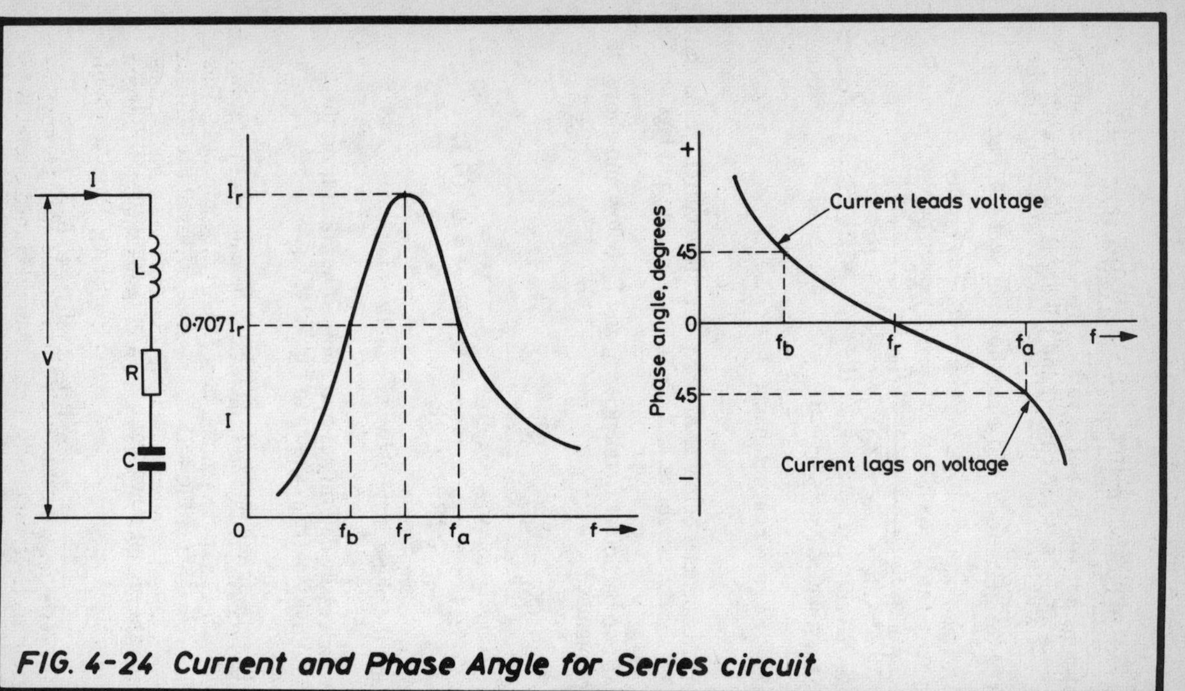

FIG. 4-24 Current and Phase Angle for Series circuit

$= 0.707 \times 0.707 = 0.5$

which in decibels is $10 \log_{10} 0.5 = -3.010$, say -3 dB.

Hence the general terms for the frequencies f_b and f_a are the lower and upper half-power frequencies respectively, or the "3 dB points".

This also happens to be a convenient level on the curve to specify the bandwidth (i.e. $f_a - f_b$), especially since it can be simply related to Q as follows:

By definition at f_b and f_a, if the appropriate angular velocities are ω_b and ω_a

$$\omega_b L - \frac{1}{\omega_b C} = R \quad \omega_a L - \frac{1}{\omega_a C} = R$$

from which

$$\omega_a - \omega_b = \frac{R}{L}$$

$$\therefore \frac{\omega_a - \omega_b}{\omega_r} = \frac{R}{\omega_r L}$$

$$\therefore \frac{f_a - f_b}{f_r} = \frac{1}{Q}$$

i.e. the Bandwidth at the 3 dB points $= \frac{f_r}{Q}$

showing simply that as Q increases, the bandwidth falls but the curve becomes steeper and hence selectivity improves. This is demonstrated in Fig.4.25 where three curves plotted through the resonant frequency of 1000 kHz are shown for various values of Q. The response scale represents the fall in current for a fixed applied voltage as the frequency deviates from f_r.

The response is most usefully shown in decibels because, in radio receivers especially, more than one tuned circuit is usually needed to provide sufficient selectivity and the rejection factors for each individual tuned circuit are then additive. The rejection factor for an unwanted signal is obtained thus:

Consider, for example a series tuned circuit with a voltage applied at f_r = 1000 kHz and Q = 100 as shown in Fig.4.25.

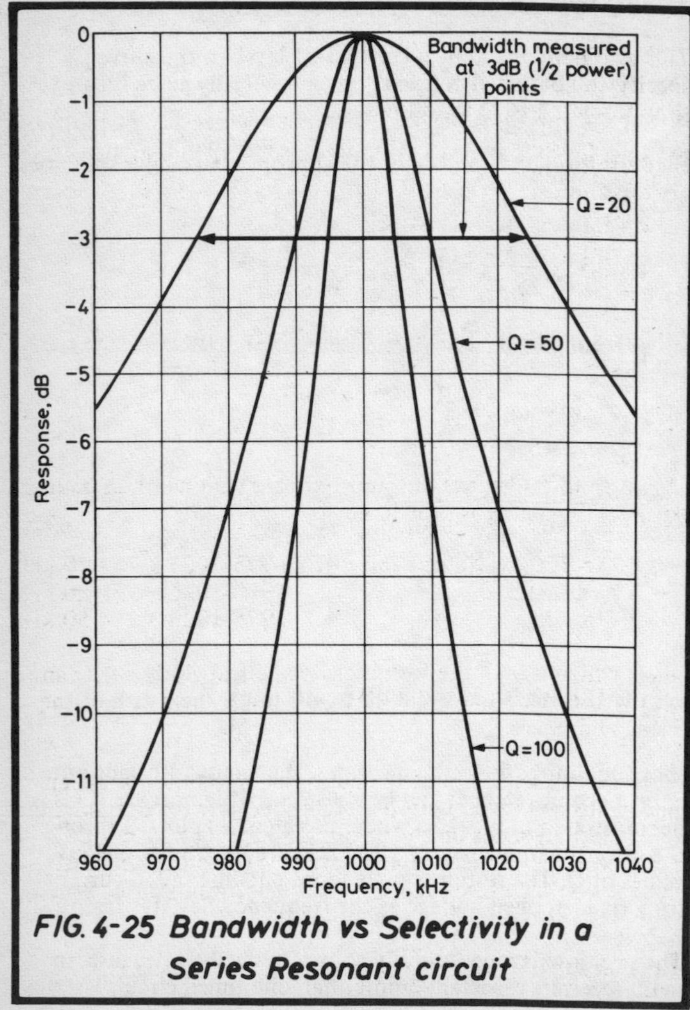

FIG. 4-25 Bandwidth vs Selectivity in a Series Resonant circuit

Let an unwanted signal appear at the same level but at a slightly different frequency, say 990 kHz. Then from the curve for Q = 100 it is evident that the current for the unwanted signal is 7dB down on that for the wanted signal, i.e. the rejection factor is 7 dB. If these two signals are fed to a second similar tuned circuit, then the output of the latter will contain the two signals but now separated by 14 dB.

Variation of Selectivity with Q

From the curves of Fig.4.25, a frequency deviating from f_r by ±10 kHz results in a circuit current lower than that at f_r by approximately

7 dB	for	Q = 100
3 dB	for	Q = 50
0.7 dB	for	Q = 20

which is evidence enough that selectivity improves as Q increases.

Variation of Bandwidth with Q

Again from the curves, the approximate bandwidth measured at the 3 dB points as shown, is

For	Q = 100, width of band is from	995-1005 kHz	= 10 kHz
	Q = 50 " " " " "	990-1010	= 20 kHz
	Q = 20 " " " " "	975-1025	= 50 kHz

all of which satisfy the formula above, bandwidth = f_r/Q and clearly the higher the value of Q, the lower the width of the pass band.

Thus the design of a tuned circuit is a compromise between bandwidth and selectivity. In a radio receiver for example, the bandwidth depends on the information being transmitted, i.e. speech, music, telegraphy, data, TV while the selectivity depends on the proximity of other adjacent unwanted signals.

Note also that $$Q = \frac{\omega_r L}{R}$$

and since $\omega_r = \dfrac{1}{\sqrt{LC}}$, $Q = \dfrac{1}{R}\sqrt{\dfrac{L}{C}}$ showing that Q is also affected by the ratio of inductance to capacitance, a further factor on which design depends.

Simplified Calculation of Resonance Curves

The full calculations involved in determining each point on a resonance curve (as in Fig.4.25) are lengthy thus a method for quick calculation is useful for the serious experimenter who wishes to determine the compromise mentioned above or examine differences in design. For very little loss of accuracy, Table 4.4 can be used to plot any curve provided that f_r and Q for the series circuit are known.

The main approximation arises from the assumption that Q does not vary over the range of frequencies near resonance. The actual variation can be shown to be very small.

Suitable methods, not requiring expensive equipment, for measurement of Q are suggested in Sect.6.2.3.

The only calculations required are to find the value of a factor d for each point to be plotted, and this is done for one side of resonance only since the other side is a mirror-image, thus

Let δ = the frequency deviation from resonance in Hz (or a multiple)

Then $d = \dfrac{\delta}{f_r} \times Q$ provided that f_r is expressed in the same multiple as δ (i.e. Hz, kHz, MHz).

e.g. at 990 kHz, with $f_r = 1000$ kHz $\delta = 1000 - 990 = 10$ kHz

$$\therefore\ d = \frac{10}{1000} \times Q = .01\,Q$$

Table 4.4 is subsequently used to obtain either the current ratio or the response relative to that at resonance in decibels, for the range of values of d.

As an example, the curves of Fig.4.25 were plotted from Table 4.4 as below:

EXAMPLE:
Plot the selectivity curve for a series tuned circuit having f_r = 1000 kHz and Q = 50.

$\frac{Q}{f_r}$ = 0.05. The following table is then constructed:

d	Current Ratio, K*	Response dB (20 $\log_{10}$ K) (all values –)	d	Current Ratio, K*	Response dB (20 $\log_{10}$ K) (all values –)
.02	0.9992	.006	0.44	0.7507	2.49
.04	0.9968	.028	0.45	0.7433	2.58
.05	0.9950	.044	0.46	0.7359	2.66
.06	0.9929	.062	0.48	0.7214	2.84
.08	0.9875	0.11	0.50	0.7071	3.01
0.1	0.9806	0.17	0.55	0.6727	3.44
0.12	0.9724	0.24	0.60	0.6402	3.87
0.14	0.9629	0.33	0.65	0.6100	4.29
0.15	0.9579	0.37	0.70	0.5812	4.71
0.16	0.9524	0.42	0.75	0.5547	5.12
0.18	0.9409	0.53	0.80	0.5300	5.51
0.20	0.9285	0.65	0.85	0.5070	5.90
0.22	0.9153	0.77	0.90	0.4856	6.28
0.24	0.9016	0.90	0.95	0.4657	6.64
0.25	0.8945	0.97	1.0	0.4472	6.99
0.26	0.8872	1.04	1.1	0.4138	7.66
0.28	0.8725	1.19	1.2	0.3846	8.30
0.30	0.8575	1.33	1.3	0.3590	8.90
0.32	0.8422	1.49	1.4	0.3363	9.47
0.34	0.8269	1.65	1.5	0.3162	10.00
0.35	0.8192	1.73	1.6	0.2983	10.51
0.36	0.8116	1.81	1.7	0.2822	10.99
0.38	0.7962	1.98	1.8	0.2676	11.45
0.40	0.7809	2.15	1.9	0.2545	11.89
0.42	0.7657	2.32	2.0	0.2425	12.31

* $\left|\frac{I}{I_r}\right|$ for Series Circuit, $\left|\frac{I_r}{I}\right|$ for Parallel Circuit.

Table 4.4 **Data for Simplified Calculation of Resonance Curves**

Col. 1 δ = Deviation from f_r, kHz	Col. 2 d = Column 1 x $\frac{Q}{f_r}$ (Column 1 x .05)	From Table 4.4	
		Current Ratio	Response, dB
0	0	1	0
2	0.1	0.98	−0.17
4	0.2	0.93	−0.65
6	0.3	0.86	−1.33
8	0.4	0.78	−2.15
10	0.5	0.71	−3.01
12	0.6	0.64	−3.87
14	0.7	0.58	−4.71
16	0.8	0.53	−5.51
18	0.9	0.49	−6.28
20	1.0	0.45	−6.99
22	1.1	0.41	−7.66
24	1.2	0.38	−8.30
26	1.3	0.36	−8.90
28	1.4	0.34	−9.47
30	1.5	0.32	−10.00
35	1.75	0.27*	−11.22*
40	2.0	0.24	−12.31

* estimated from d = 1.7 and 1.8

Both sides of the selectivity curve (with the vertical axis marked in decibels or current ratio, or both as required) can then be plotted as in Fig.4.25.

EXAMPLE:
A radio station broadcasts on a frequency of 908 kHz. What is the minimum Q for a series tuned circuit to reject an unwanted signal at 903.5 kHz by 10 dB?

From Table 4.4, for 10 dB rejection, d = 1.5

and since $d = \frac{\delta . Q}{f_r}$ i.e. $Q = \frac{f_r d}{\delta}$

$$\therefore Q = \frac{908 \times 1.5}{4.5} = 303$$

Parallel Circuits

Most of the considerations with regard to the selectivity and bandwidth of parallel tuned circuits follow those for the series circuit but in a reciprocal manner, this is illustrated by the fact that at resonance the series circuit has minimum impedance with maximum current, whereas the parallel case produces maximum impedance and minimum current. Generally the usefulness of the former lies in the resonant rise in voltage across the components whereas that of the latter comes from development of voltage across the high impedance of the circuit.

Continuing on the "reciprocal" theme, it will be found that if $|I_r/I|$ is plotted as a resonance curve for a parallel circuit, the same relationships as for the series circuit arise, e.g. that selectivity varies with Q and also, if f_b and f_a are the two frequencies where $I = I_r/.707$ (the current has *risen* from the resonant value I_r to $\sqrt{2}I_r$), the bandwidth $f_a - f_b$ is again equal to f_r/Q.

Thus it is legitimate to use Table 4.4 to determine the selectivity of a parallel circuit provided that the current ratio used is $|I_r/I|$ or equally the scale could be $|Z/Z_r|$ (off resonance, the impedance is falling).

EXAMPLE:
A parallel tuned circuit (as in Fig.4.21) comprises an inductor L = 3 mH, capacitor, C = 100 pF and resistance R = 137 ohms the circuit resonating at 290.6 kHz. What is the value of the impedance (modulus only) at 10 kHz above the resonant frequency?

$$Z_r = \frac{L}{CR} = \frac{3 \times 10^{-3}}{100 \times 10^{-12} \times 137} = 219 \times 10^3 \text{ ohms,}$$

$$Q_r = \frac{\omega L}{R} = \frac{2\pi \times 290.6 \times 10^3 \times 3 \times 10^{-3}}{137} \simeq 40$$

$$d = \frac{\delta Q}{f_r} = \frac{10 \times 40}{290.6} = 1.38$$

From Table 4.4 for d = 1.38, $|I_r/I|$ or $|Z/Z_r| \simeq 0.34$

i.e. Impedance at 10 kHz above resonance = $0.34 \times Z_r$
= 74460 ohms

(Note that in this case, for a sacrifice of less than 0.7% of accuracy, the calculations arising from the formula

$$|Z| = \frac{1}{\sqrt{\left(\frac{R}{R^2+\omega^2L^2}\right)^2 + \left(\frac{\omega L}{R^2+\omega^2L^2} - \omega C\right)^2}} \quad \text{(Sect.4.5.}$$

have been avoided.)

4.7.5 Coupled Circuits

Two circuits may be coupled together magnetically, the coupling component being known as a transformer. However, because in general transformers are not resonant devices, they are considered at greater length in Sec.3.5 – in this section only those arrangements are considered in which the transformer windings are part of high-frequency, loosely-coupled resonant circuits as, for example, in the case of an intermediate-frequency transformer.

Consider the circuit of Fig.4.26 which shows two circuits P and S magnetically coupled via the inductive windings L_p and L_s.

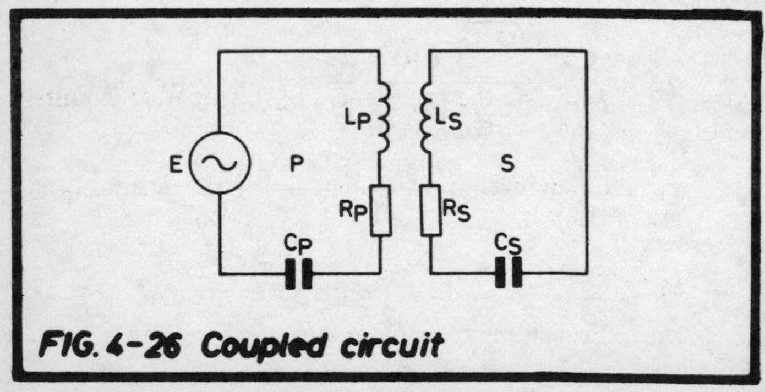

FIG. 4-26 Coupled circuit

The fact that a current in one circuit induces a voltage in the other is expressed by a quantity known as the mutual inductance (M) which theoretically can range from zero when the two windings are completely separated magnetically, to maximum when the coupling is perfect, whence

$$M_{max} = \sqrt{L_p L_s}\,.$$

To express the degree of coupling between these two limits, a "coefficient of coupling", k, is used, this is defined as the ratio between the mutual inductance in any particular case to the maximum value i.e.

$$k = \frac{M}{\sqrt{L_p L_s}}$$

"Critical coupling" is obtained with a particular value of k (= k_c) at which

(i) secondary current is maximum

(ii) the impedance coupled into the primary by the secondary is equal to the resistance of the primary i.e.

$$\frac{(\omega M)^2}{R_s} = R_p \qquad \text{(see below)}$$

(iii) from the latter expression it can be shown that

$$k_c = \frac{1}{\sqrt{Q_p Q_s}}$$

Also, if Z_p and Z_s are the impedances of the separate circuits and I_p and I_s their currents, then

(i) the e.m.f. induced in the secondary by the current I_p in the primary is

$$e_s = -j\omega M I_p$$

(the $-j$ indicating that this voltage lags on I_p by 90°)

(ii) the effect of the secondary circuit on the primary circuit is to couple an impedance $(\omega M)^2/Z_s$ in series into the primary and similarly from primary to secondary, adding in series $(\omega M)^2/Z_p$.

Considering both circuits being tuned to the same frequency, if k is small, then I_s is small and the response/frequency curve peaks singly at f_r (Fig.4.27). As k is increased to its critical value, I_s reaches maximum and becomes slightly flat-topped. A further increase in k causes "double-humping" which, with careful design can produce a characteristic having a reasonably even response over a band of frequencies with good discrimination against unwanted signals outside of this band, i.e. a band-pass filter.

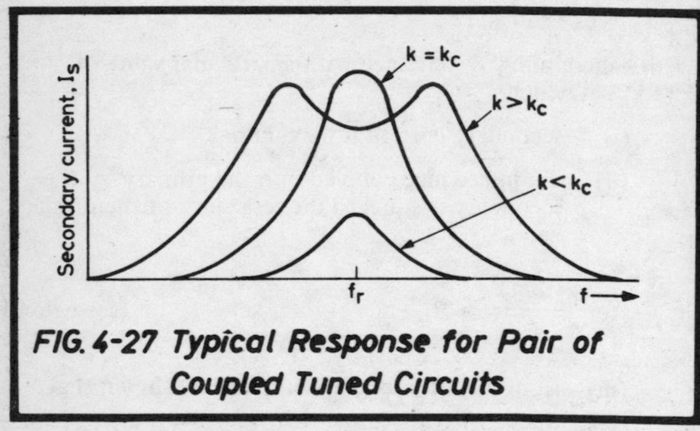

FIG. 4-27 Typical Response for Pair of Coupled Tuned Circuits

With even further increase of k, the width of the pass-band between the two peaks increases and if f_1 and f_2 are the frequencies at which the two peaks occur, then approximately

$$\text{width of pass band} = f_r \; \frac{1}{\sqrt{1-k}} - \frac{1}{\sqrt{1+k}}$$

$$\text{since the higher of the two frequencies, } f_2 = \frac{f_r}{\sqrt{1-k}}$$

$$\text{and also } f_1 = \frac{f_r}{\sqrt{1+k}}$$

Over the range considered above, k is still small and thus

$$\text{width of pass band} \simeq kf_r.$$

In practice, concern is mainly with selectivity and bandwidth and although production of a response/frequency curve is more difficult than for a single tuned circuit (Sect.4.7.4) it is relatively easy to estimate the curve when both primary and secondary circuits are tuned to the same frequency. The principle involved is that simplified equations may be used to determine the turning points of the secondary current characteristic as in Fig.4.28, i.e. the points at frequencies f_1, f_r and f_2, and in addition, the two frequencies below and above f_r (f_3 and f_4) at which the secondary current is equal to its value at resonance are easily calculated. This helps to estimate the slope of the characteristic at these frequencies, which is important in determining the degree of rejection of unwanted signals.

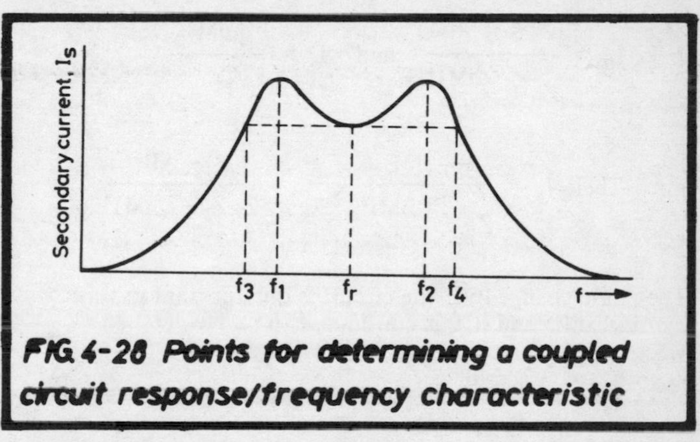

FIG.4-28 Points for determining a coupled circuit response/frequency characteristic

Note that the smaller $(f_1 - f_3)$ and $(f_4 - f_2)$ become, the greater the discrimination.

This is best explained by an example, followed by a list of steps to be taken.

EXAMPLE:
The circuit and component values of an IF transformer are given in Fig.4.29. Sketch the response/frequency characteristic.

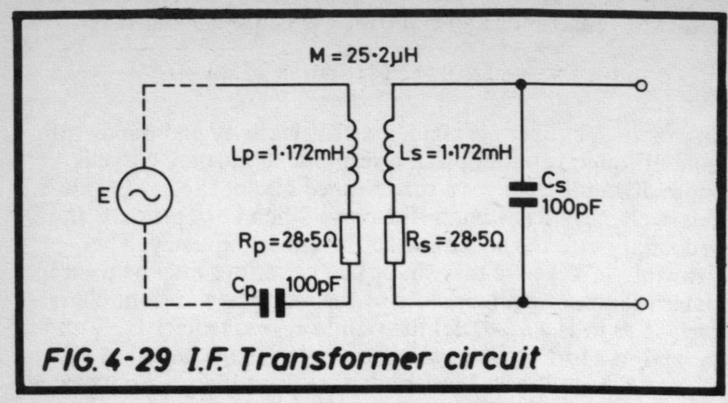

FIG. 4-29 I.F. Transformer circuit

Current, I_s at f_r:

Since, from the rules given above,

$$I_p = \frac{E}{Z_p + \dfrac{(\omega M)^2}{Z_s}} \quad \text{and} \quad I_s = \frac{-j\omega M I_p}{Z_s}$$

$$\text{Then } I_s = \frac{-j\omega ME}{Z_p + \dfrac{(\omega M)^2}{Z_s}} \times \frac{1}{Z_s} = \frac{-j\omega ME}{Z_p Z_s + (\omega M)^2}$$

The phase change over the circuit is unimportant in this consideration and if E is considered to be 1 volt, then

$$I_s = \frac{\omega M}{Z_p Z_s + (\omega M)^2} \quad \text{and at resonance} = \frac{\omega_r M}{R_p R_s + (\omega_r M)}$$

Substituting the values from Fig.4.29

$$\omega_r = \frac{1}{\sqrt{LC}} = \frac{1}{\sqrt{1.172 \times 10^{-3} \times 100 \times 10^{-12}}} = \frac{10^7}{\sqrt{11.72}}$$

$$= 2921070 \text{ rads/sec}$$

$$\simeq 2921 \times 10^3 \text{ rads/sec}$$

$$\therefore I_s \text{ at } f_r = \frac{2921 \times 10^3 \times 25.2 \times 10^{-6}}{28.5 \times 28.5 + (2921 \times 10^3 \times 25.2 \times 10^{-6})^2}$$

$$= \frac{73.61}{812.25 + (73.61)^2}$$

$$= 11.81 \text{ mA}$$

$$k = \frac{M}{\sqrt{L_p L_s}} = \frac{25.2 \times 10^{-6}}{1.172 \times 10^{-3}} = .0215$$

$$f_r = \frac{\omega_r}{2\pi} = \frac{2921 \times 10^3}{2\pi} \simeq 465 \text{ kHz}$$

$$f_1 = \frac{f_r}{\sqrt{1+k}} \qquad f_2 = \frac{f_r}{\sqrt{1-k}}$$

$$= \frac{465}{\sqrt{1+.0215}} \text{ kHz} \qquad = \frac{465}{\sqrt{1-.0215}} \text{ kHz}$$

$$= 460.08 \text{ kHz} \qquad = 470.08 \text{ kHz}$$

(note that f_r is not exactly midway between f_1 and f_2).

Current at f_1 and f_2

It can be shown that the secondary current at f_1 and f_2 is approximately equal to the maximum resonance value (which occurs with critical coupling)

$$\therefore I_s \text{ at } f_1 \text{ and } f_2 = \frac{\omega_r M}{R_p R_s + (\omega_r M)^2} \qquad \text{But } (\omega_r M)^2 = R_p R_s$$

$$\therefore I_s \text{ at } f_1 \text{ and } f_2 = \frac{1}{2\sqrt{R_p R_s}} = \frac{1}{2 \times 28.5} = 17.54 \text{ mA}$$

Frequencies for secondary current equal to resonance value (f_3 and f_4)

The two frequencies above and below f_r at which the secondary current is equal to its resonance value are given approximately by

$$f_r \pm .707\,(f_2 - f_1)$$

$$= f_r \pm 7.07 \text{ kHz}$$

From this information the characteristic can be sketched as in Fig.4.30.

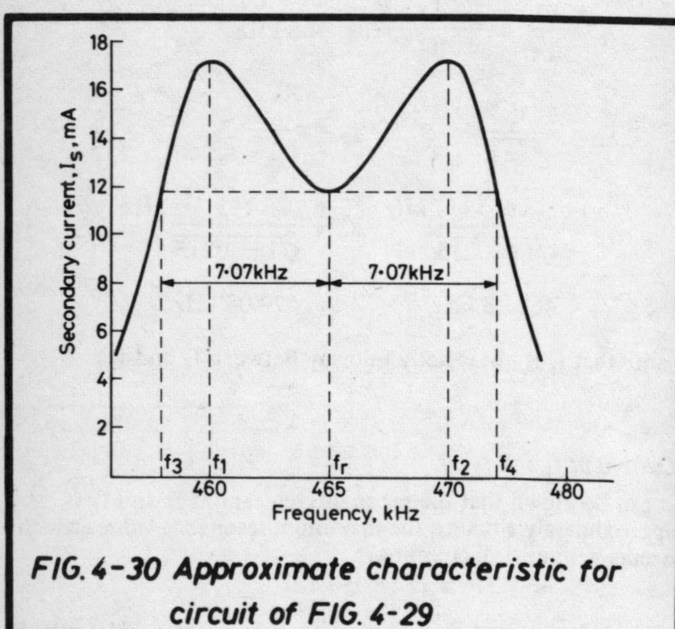

FIG. 4-30 Approximate characteristic for circuit of FIG. 4-29

Summary of Method

Calculate:

1. $\omega_r \quad = \dfrac{1}{\sqrt{LC}}$

2. $f_r = \dfrac{\omega_r}{2\pi}$

3. $\omega_r M$

4. $(\omega_r M)^2$

5. $k = \dfrac{M}{\sqrt{L_p L_s}}$. For identical primary and secondary, = M/L

6. f_1 and f_2 = $\dfrac{f_r}{\sqrt{1+k}}$ and $\dfrac{f_r}{\sqrt{1-k}}$

Then

$$\text{Current at } f_r = \frac{\omega_r M}{R_p R_s + (\omega_r M)^2}$$

$$\text{Current at } f_1 \text{ and } f_2 = \frac{1}{2\sqrt{R_p R_s}}$$. For identical primary and secondary, = 1/2R

Frequencies for Secondary Current equal to resonance value (f_3 and f_4)

$$f_3 = f_r - .707\,(f_2 - f_1) \quad \text{and} \quad f_4 = f_r + .707\,(f_2 - f_1).$$

There may be doubt that the characteristic will exhibit two peaks (the double-hump). In this case, k_c must also be calculated and compared with k. As already explained, two peaks only appear when k exceeds k_c.

4.8 CALCULATION OF POWER

Power, in the direct current case, as shown in Sect.2.1 is calculated by

$$V \times I = \frac{V^2}{R} = I^2 R \text{ watts}$$

where V is in volts,
I is in ampères,
R is in ohms.

In the alternating current case, the laws equally apply at any instant, i.e. the instantaneous power being dissipated is obtained by multiplying the voltage and current together at that instant. This must be stated because voltage and current in an a.c. circuit are not necessarily in phase. In fact, if the voltage and current waveforms in purely inductive or capacitive circuits are considered together, it is found that the phasor dissipated over each half-cycle is zero.

Usually an a.c. circuit has some phase angle, ϕ, between voltage and current represented typically in the phasor diagram of Fig.4.31.

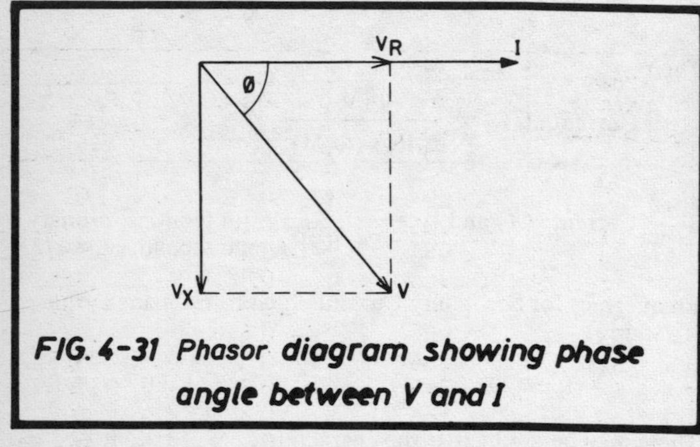

FIG. 4-31 Phasor diagram showing phase angle between V and I

The reference phasor is the current I, on which is marked off the voltage V_R developed across the circuit resistance and at 90°, the voltage V_X across the net circuit reactance. Then the applied voltage is the phasor V, lagging (in this example) by an angle ϕ.

The power is calculated by multiplying I by the component of V in phase with it, i.e. V_R and since $V_R = V \cos \phi$

$$\text{Power} = VI \cos \phi$$

The term $\cos \phi$ is known as the "power factor" and is the factor by which the apparent power (also referred to as 'volt-amps') given by $V \times I$ is multiplied to obtain the true power. Hence

$$\text{Power Factor } (\cos\phi) = \frac{\text{True Power}}{\text{Apparent Power}}$$

and because the impedance of the circuit is expressed in the form $Z\angle\phi$ then

$$\cos\phi = \frac{\text{Resistance}}{\text{Impedance}}$$

Table 4.5 gives the values of $\cos\phi$ to three significant figures at 5° and 0.1 radian intervals. Other phase angle values may be obtained from any book of standard tables.

Phase Angle, ϕ Degrees	Power Factor, $\cos\phi$	Phase Angle, Radians	Power Factor, $\cos \frac{360}{2\pi}$ rads.
0	1.00	0	1.00
5	.996	0.1	.995
10	.985	0.2	.980
15	.966	0.3	.955
20	.940	0.4	.921
25	.906	0.5	.878
30	.866	0.6	.825
35	.819	0.7	.765
40	.766	0.8	.697
45	.707	0.9	.622
50	.643	1.0	.540
55	.574	1.1	.454
60	.500	1.2	.363
65	.423	1.3	.268
70	.342	1.4	.170
75	.259	1.5	.071
80	.174	1.571	0
85	.087		
90	0		

Table 4.5 **Conversion of Phase Angle to Power Factor**

4.9 SKIN EFFECT

When a conductor carries an alternating current, the changing magnetic flux (Sect.2.3.1) causes the effective resistance of the conductor to increase.

The flux is concentric with the conductor and whereas the centre of the conductor is linked by all the flux, the surface (or skin) is linked only by the flux external to the conductor (hence the term "skin effect").

In terms of inductance, the conductor has greater self-inductance (and therefore reactance) at the centre than at the surface, equally the effect can be explained in terms of back-e.m.f. which is also greater at the centre than at the surface.

The result is that the current is constrained to flow more along the surface than in the centre, effectively resulting in a loss of conductor cross-sectional area and therefore a rise in resistance.

Skin effect is usually expressed as the ratio Rac/Rdc for the particular conductor at a given frequency, f. The value of the ratio varies in a complex manner with (i) conductor conductivity, (ii) conductor diameter or cross-sectional area, (iii) frequency. For any particular conductor Rac/Rdc $\propto \sqrt{f}$.

If copper conductors only are considered, then (i) is constant and (ii) and (iii) may be expressed in the form $d\sqrt{f}$ (where d = conductor diameter) and values of Rac/Rdc may be obtained for a practical range of values of $d\sqrt{f}$. This is displayed in Table 4.6 where Rac/Rdc is expressed directly from combinations of d and f.

Although figures will not apply exactly in any given case they are useful to show the order of change of 'high frequency resistance' when values of Q are being calculated. Also when the conductor is wound into a coil, skin effect becomes even more complicated and greater in magnitude.

EXAMPLE:
A 10 m length of copper wire of diameter 0.9 mm has a resistance of 0.271 ohms. What will be its effective resistance when carrying a current at 4 MHz?

f / d/mm	kHz →									MHz →									
	100	200	300	400	500	600	700	800	900	1	2	3	4	5	6	7	8	9	10
0.2	1.002	1.005	1.008	1.015	1.025	1.035	1.05	1.065	1.08	1.18	1.32	1.56	1.77	1.95	2.11	2.26	2.40	2.53	2.65
0.3	1.005	1.018	1.045	1.08	1.12	1.17	1.21	1.26	1.33	1.38	1.86	2.22	2.53	2.80	3.04	3.27	3.48	3.67	3.86
0.4	1.015	1.06	1.135	1.29	1.32	1.42	1.52	1.61	1.69	1.76	2.40	2.88	3.29	3.65	3.98	4.28	4.56	4.82	5.07
0.5	1.04	1.15	1.28	1.45	1.59	1.72	1.84	1.95	2.05	2.15	2.94	3.54	4.05	4.51	4.91	5.29	5.63	5.96	6.27
0.6	1.08	1.26	1.49	1.69	1.86	2.01	2.16	2.29	2.41	2.53	3.48	4.20	4.82	5.36	5.85	6.30	6.71	7.11	7.48
0.7	1.14	1.43	1.70	1.93	2.13	2.31	2.48	2.63	2.77	2.91	4.01	4.86	5.58	6.21	6.78	7.31	7.79	8.25	8.68
0.8	1.29	1.60	1.91	2.17	2.40	2.60	2.80	2.97	3.14	3.29	4.55	5.53	6.34	7.06	7.72	8.31	8.87	9.39	9.89
0.9	1.33	1.78	2.12	2.41	2.67	2.90	3.12	3.31	3.50	3.67	5.09	6.19	7.11	7.92	8.65	9.32	9.95	10.5	11.1
1.0	1.45	1.95	2.33	2.65	2.94	3.20	3.44	3.65	3.86	4.05	5.63	6.85	7.87	8.77	9.58	10.3	11.0	11.7	12.3
1.1	1.57	2.12	2.54	2.89	3.21	3.49	3.76	4.00	4.22	4.43	6.17	7.51	8.63	9.62	10.5	11.3	12.1	12.8	13.5
1.2	1.69	2.29	2.75	3.14	3.48	3.79	4.08	4.34	4.58	4.81	6.71	8.17	9.39	10.5	11.5	12.4	13.2	14.0	14.7
1.3	1.81	2.46	2.96	3.38	3.75	4.08	4.40	4.68	4.94	5.20	7.25	8.83	10.2	11.3	12.4	13.4	14.3	15.1	15.9
1.4	1.93	2.63	3.17	3.62	4.02	4.38	4.72	5.02	5.31	5.58	7.79	9.49	10.9	12.2	13.3	14.4	15.3	16.3	17.1
1.5	2.05	2.80	3.37	3.86	4.29	4.67	5.03	5.36	5.67	5.96	8.33	10.2	11.7	13.0	14.3	15.4	16.4	17.4	18.3
1.6	2.17	2.97	3.58	4.10	4.56	4.97	5.35	5.70	6.03	6.34	8.87	10.8	12.5	13.9	15.2	16.4	17.5	18.6	19.5
1.7	2.29	3.14	3.79	4.34	4.82	5.26	5.67	6.04	6.39	6.72	9.41	11.5	13.2	14.7	16.1	17.4	18.6	19.7	20.8
1.8	2.41	3.31	4.00	4.58	5.10	5.56	5.99	6.38	6.75	7.10	9.95	12.1	14.0	15.6	17.1	18.4	19.7	20.8	22.0
1.9	2.53	3.48	4.21	4.82	5.34	5.85	6.31	6.73	7.12	7.48	10.5	12.8	14.7	16.5	18.0	19.4	20.7	22.0	23.2
2.0	2.65	3.65	4.42	5.07	5.63	6.15	6.63	7.07	7.48	7.86	11.0	13.5	15.5	17.3	18.9	20.4	21.8	23.1	24.4

Table 4.6 **Rac/Rdc for Circular Cross-Section Solid Copper Wires**

$\frac{Rac}{Rdc}$ is obtained directly from Table 4.6 i.e. 7.1

$\therefore$ Effective resistance of wire at 4 MHz $= 0.271 \times 7.1$
$= 1.924$ ohms

5. NETWORKS AND THEOREMS

5.1 NETWORK ANALYSIS

There are several theorems and laws which greatly facilitate the analysis of electrical networks, the most widely known of these are set out in this section.

5.1.1 Kirchhoff's Laws

These enable simultaneous equations to be derived for calculating the values of currents in a network.

Law 1

The algebraic sum of the currents meeting at a point in a network is zero.

Law 2

In any closed circuit (or "mesh") the algebraic sum of the e.m.f.'s is equal to the algebraic sum of the products of the resistances and the respective currents in the separate parts.

Both laws can be demonstrated by means of the simple (i.e. resistances only) Wheatstone Bridge network of Fig.5.1.

Considering point A, then from Law 1 which in effect states that the current entering a point is equal to the current leaving it,

$I_1 = I_2 + I_3$ so that for I_3 could be written $(I_1 - I_2)$,

eliminating one unknown quantity.

Again, if I_2 flows into point D and I_4 and I_5 flow away, then

$I_5 = I_2 - I_4$ or $I_4 = I_2 - I_5$

Law 2 is illustrated by consideration of any closed circuit or mesh e.g. battery E, R_2 and R_4.

Then $E = I_2 R_2 + I_5 R_4$

Similarly for the mesh ADBA,

$0 = I_2 R_2 + I_4 Rg - I_3 R_1$ (note that the p.d. $I_3 R_1$ is in opposition to the other two).

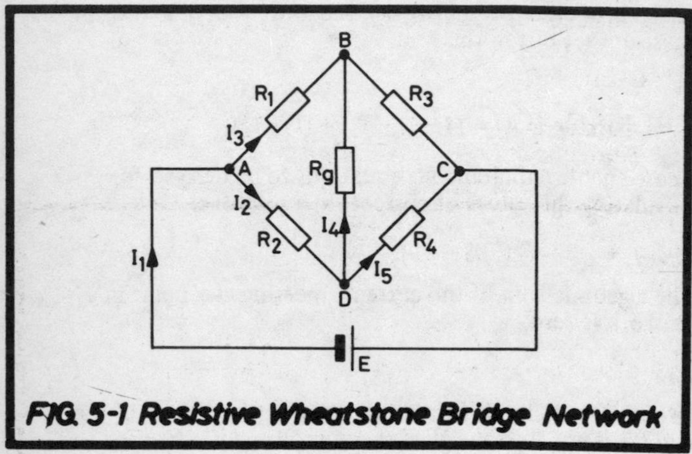

FIG. 5-1 Resistive Wheatstone Bridge Network

EXAMPLE:
Find the current in each resistor in the network shown in Fig.5.2.

Because some of the meshes considered using Law 2 will have common branches, it is important first to establish a direction of current flow which is considered to be, say, positive. In this example we choose the conventional current direction which is anticlockwise from the battery but we could equally work on the basis of electron flow (sect. 2.3.1) and go clockwise.

From Law 1: Considering point C

Current in $= I$
Current out $= I_1 + (I - I_1)$

Considering point B, if I_2 flows from B to D, (the true direction is unknown at this stage)

$$\text{Current in} = (I - I_1)$$
$$\text{Current out} = (I_2 + I_3) \qquad \text{but } I_3 = (I - I_1 - I_2)$$
$$\therefore \quad \text{"} \quad \text{"} = I_2 + (I - I_1 - I_2)$$

Considering point D

$$\text{Current in} = I_1 + I_2$$
$$\text{Current out (i.e. D to A)} = I_1 + I_2$$

Check at point A

$$\text{Current in} = (I - I_1 - I_2) + (I_1 + I_2) = I$$
$$\text{Current out} = I$$

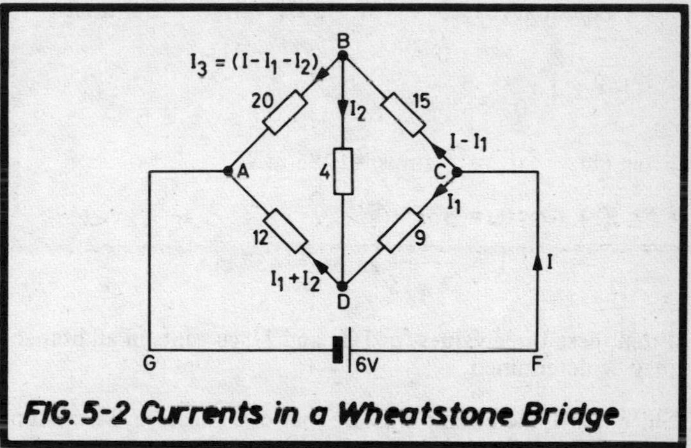

FIG. 5-2 Currents in a Wheatstone Bridge

Currents in each of the five resistors have therefore been allocated, using three unknowns, for which three simultaneous equations are required for complete solution, i.e. three separate meshes must be considered.

From Law 2: Mesh AGFCDA

$$6 = 9I_1 + 12(I_1 + I_2)$$
$$\therefore 2 = 7I_1 + 4I_2 \qquad (1)$$

Mesh DCBD

$0 = 15(I - I_1) + 4I_2 - 9I_1$ (I_1 is clockwise, therefore negative according to the convention adopted)

$\therefore 0 = 15I - 24I_1 + 4I_2$ (2)

Mesh ADBA

$0 = -4I_2 - 12(I_1 + I_2) + 20(I - I_1 - I_2)$

$\therefore 0 = 5I - 8I_1 - 9I_2$ (3)

Solution:

Multiply equation (3) by 3

$0 = 15I - 24I_1 - 27I_2$

Equation (2) $0 = 15I - 24I_1 + 4I_2$ Subtract

$0 = -31I_2$

$\therefore I_2 = 0$

From (1) $I_1 = \frac{2}{7}$ amps $= 286$ mA

" (2) $0 = 15I - \frac{48}{7}$

$$\therefore I = \frac{48}{7 \times 15} \text{ amps} = 457 \text{ mA}$$

From these three values for I, I_1 and I_2, currents in all branches may be determined.

Since $I_2 = 0$, the circuit must be that of a *balanced* Wheatstone Bridge.

5.1.2 Wheatstone Bridge

The unbalanced bridge circuit can be solved completely by use of Kirchhoff's Laws but the most useful conclusions reached from the bridge are when it is in balance. Consider Fig.5.1. For zero current in Rg, B and D must be at the same potential

i.e. p.d. across R_1 = p.d. across R_2

$$\therefore I_3R_1 = I_2R_2 \qquad \therefore \frac{I_2}{I_3} = \frac{R_1}{R_2}$$

Also because the current in R_3 is I_3 and in R_4 is I_2 (no current is bypassed through Rg)

$$I_3 R_3 = I_2 R_4 \qquad \therefore \qquad \frac{I_2}{I_3} = \frac{R_3}{R_4}$$

Hence, at balance $\dfrac{R_1}{R_2} = \dfrac{R_3}{R_4}$, also $\dfrac{R_1}{R_3} = \dfrac{R_2}{R_4}$

In the general case, some or all arms will be impedances as in Fig.5.3, E being an a.c. generator.

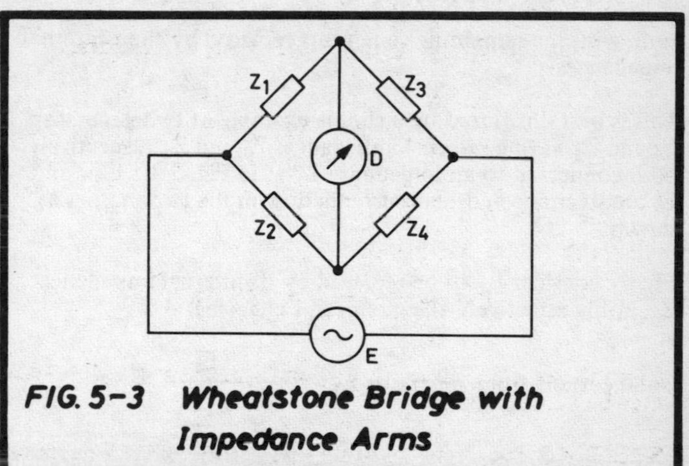

FIG. 5-3 Wheatstone Bridge with Impedance Arms

Then for zero deflexion on detector D (or zero tone in receiver)

$$\frac{Z_1}{Z_2} = \frac{Z_3}{Z_4} \quad \text{and} \quad \frac{Z_1}{Z_3} = \frac{Z_2}{Z_4}$$

$$\therefore \quad \frac{|Z_1| \angle \phi_1}{|Z_3| \angle \phi_3} = \frac{|Z_2| \angle \phi_2}{|Z_4| \angle \phi_4}$$

which is satisfied by the two conditions $|Z_1|/|Z_3 = |Z_2|/|Z_4|$ and

$\phi_1 - \phi_3 = \phi_2 - \phi_4$ i.e. for balance, the impedance moduli must be in the correct ratio and the net phase angle differences on both sides of the detector must be equal.

Examples of the use of the Wheatstone Bridge are in Section 6.

5.1.3 Superposition Theorem

In a network of linear impedances (i.e. impedances for which Ohm's Law applies) containing more than one generator, the current at any point is the vector sum of the individual currents that would flow if each generator were considered in turn, with the remaining generators replaced by their internal impedances.

This is best illustrated by a simple example of two generators E_1 and E_2 having internal impedances Z_1 and Z_2 respectively, both connected to an impedance Z_3 as in Fig. 5.4(a). E_1 and E_2 are so connected that currents flow in the two meshes as shown.

Firstly consider E_1 to be replaced by its internal impedance Z_1, giving effectively the circuit in Fig.5.4(b).

$$\text{Total current from generator } E_2 = \frac{E_2}{Z_2 + \dfrac{Z_1 Z_3}{Z_1 + Z_3}} = I$$

$$\text{and current } I_3 \text{ through } Z_3 = I \times \frac{Z_1}{Z_1 + Z_3}$$

$$= \frac{E_2 Z_1}{Z_1 Z_2 + Z_1 Z_3 + Z_2 Z_3}$$

i.e. the current in Z_3 due to E_2 alone but taking into account the impedance of generator E_1.

Similarly, if E_2 is replaced by its internal impedance Z_2

$$\text{Current in } Z_3 = \frac{E_1 Z_2}{Z_1 Z_2 + Z_1 Z_3 + Z_2 Z_3}$$

and total current due to both generators is the sum of the two currents calculated above,

$$\text{i.e.} \quad \frac{E_1 Z_2 + E_2 Z_1}{Z_1 Z_2 + Z_1 Z_3 + Z_2 Z_3}$$

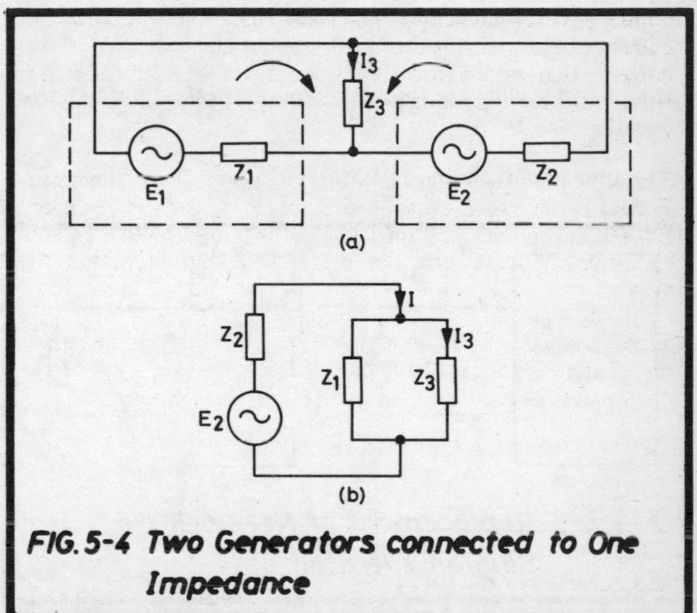

FIG. 5-4 Two Generators connected to One Impedance

Solving this particular network by use of the Superposition Theorem is in fact very little easier than by using Kirchhoff's Laws. However, as the complexity of the circuit increases, the value of the Superposition Theorem usually predominates owing to the greater difficulty of handling many simultaneous equations with the Kirchhoff method.

5.1.4 Thévenin's Theorem

This is better explained diagrammatically but the full theorem is first given.

The current in a load impedance connected to the output terminals of a network of impedances and generators is unchanged if the network is replaced by a generator having (i) a constant e.m.f. equal to the open-circuit voltage measured looking back into the output terminals of the network, (ii) an impedance equal to that looking back into the output terminals with each generator replaced by its own internal impedance.

As in the preceding section, the theorem applies only to linear impedances.

Diagrammatically, as in Fig.5.5(a).

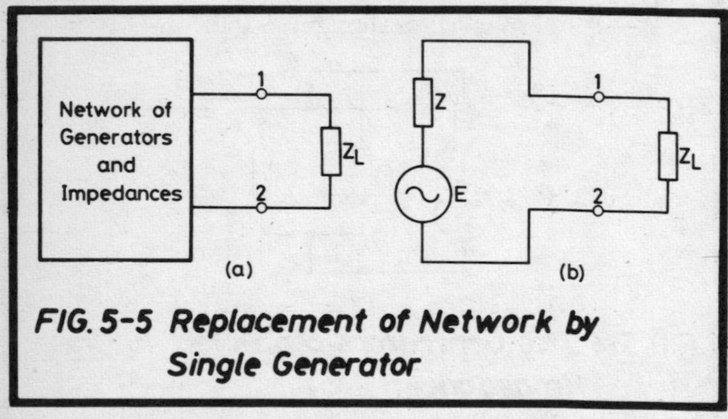

FIG. 5-5 Replacement of Network by Single Generator

The load is disconnected, then

(1) The open-circuit voltage appearing across terminals 1 and 2 is measured (E).

(2) Each generator is replaced by its own internal impedance and the network impedance measured at terminals 1 and 2 (Z).

The network of Fig.5.5(a) can then be simulated by a generator of e.m.f. E and impedance Z as in Fig.5.5(b).

In practice, although measurement of (1) above is straightforward, carrying out (2) to obtain Z is not always practicable (e.g. the generators may be transistors). The following example shows how this is avoided.

EXAMPLE:
Suppose a network (e.g. a power amplifier) has two output terminals 1 and 2 and the open-circuit voltage measured across them is 3 volts at a certain frequency. Two further measurements show that (i) a current of 35 mA flows between terminals 1 and 2 when they are short-circuited (ii) a current of 28.7 mA flows through a load of 20 ohms.

Find the equivalent circuit and the current which will flow in an 8 ohm load.

E of the equivalent circuit is given, i.e. 3 volts.

Let $Z = \sqrt{R^2 + X^2}$, then from (i) $.035 = \dfrac{3}{\sqrt{R^2 + X^2}}$

$$\therefore\ R^2 + X^2 = \frac{3^2}{.035^2} = 7347 \text{ ohms}$$

from (ii) $$.0287 = \frac{3}{\sqrt{(R+20)^2 + X^2}}$$

$$\therefore\ (R+20)^2 + X^2 = \frac{3^2}{(.0287)^2} = 10926 \text{ ohms}$$

$$\therefore\ R^2 + 40R + 400 + X^2 = 10926$$

$$\therefore\ R^2 + 40R + X^2 = 10526$$

also $$R^2 + X^2 = 7347$$

$$40R = 3179$$

$$\therefore\ R = 79.475 \text{ ohms}$$

and since $R^2 + X^2 = 7347$

$$X^2 = 7347 - 79.475^2 = 1030.7$$

$$X = 32.1 \text{ ohms}$$

i.e. the equivalent circuit is that of a generator of 3V with impedance 79.475 + j32.1 ohms.

$$\text{Current in 8 ohm load} = \frac{3}{\sqrt{(79.475 + 8)^2 + 32.1^2}}$$

$$= \frac{3}{93.18} \text{ amp} = 32.2 \text{ mA}$$

5.1.5 Maximum Power Transfer Theorem

An important theorem mainly used to provide the guiding principles for interconnecting networks with lines, aerials, transducers etc.

The full theorem states that

(1) The maximum power will be obtained from a generator of internal impedance $Z\angle\phi$ if its load has the conjugate impedance $Z\angle-\phi$. In the form $R + jX$ for the generator impedance, maximum power transfer occurs with a load impedance $R - jX$.

(2) If the modulus only can be varied, the power will be maximum when the moduli of generator and load are equal, irrespective of the value of ϕ.

Transference of maximum power into a load from a generator should not be confused with maximum efficiency. When a generator is matched to its load, the efficiency is 50%. Higher efficiencies are obtained as load impedance is increased above the matched value but the power into the load actually decreases. High voltage power-line transformers are not matching transformers, they are more concerned with efficiency because power losses cost money e.g. in this case running at higher voltages and therefore lower currents to reduce I^2R losses in the lines.

Frequently transformers are used for impedance matching, e.g. matching a loudspeaker to the output of an amplifier (the use of transformers for this particular purpose is considered more fully in Sect.3.5) or matching a line to the input or output of a network. The transmission loss of a transformer is low. Passive networks not employing transformers may also be used for matching (Sect.5.3) but in this case the network itself introduces a loss.

The use of a transformer arises from the fact that its impedance ratio is equal to the square of its turns ratio so that if Z_L is a load impedance connected to the secondary of a matching transformer of turns ratio N_P/N_S, where N_P and N_S are the turns on the appropriate windings, then the modulus of the impedance reflected back into the primary is

$$Z_L\left(\frac{N_P}{N_S}\right)^2 .$$

The angle of the impedance reflected back is unchanged, hence the second part of the theorem which caters for the more practical arrangement, i.e. perfect matching is seldom obtained because of the added complication of making the impedance angles equal and of opposite sign.

Consider the circuit in Fig.5.6 which shows the matching required from, say, a coaxial line to an amplifier input.

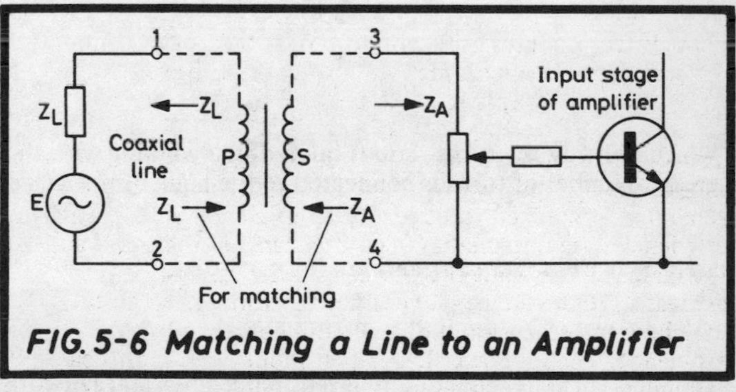

FIG. 5-6 Matching a Line to an Amplifier

By Thévenin's Theorem, the line may be represented by a generator of e.m.f. E and internal impedance Z_L equal to that of the characteristic impedance of the line.

Suppose a transformer to be connected as shown dotted having a turns ratio

$$T = \frac{N_P}{N_S} \quad \text{so that} \quad T^2 = \frac{|Z_L|}{|Z_A|}.$$

Then impedance seen by line $= |Z_A| \times T^2$, but $|Z_A| = \dfrac{|Z_L|}{T^2}$

$\therefore$ " " " " $= |Z_L|$

Similarly the impedance seen by the amplifier is $|Z_A|$, thus giving the condition of optimum matching.

EXAMPLE:
Suppose in the case of Fig.5.6 the coaxial line has an impedance of 75 ohms and the amplifier input 550 ohms (both resistive). What transformer turns ratio will give optimum matching?

$$Z_L = 75 \text{ ohms}, \quad Z_A = 550 \text{ ohms}.$$

Impedance seen by line = $550 \times T^2$ which must also equal Z_L

$$\therefore 75 = 550T^2 \quad T = \frac{75}{\sqrt{550}} = 0.369{:}1$$

or equally $\dfrac{1}{0.369} = 2.71{:}1$

Whichever way the turns ratio is quoted, the winding with the greater number of turns is connected to the higher impedance.

5.1.6 Star-Delta Transformation

(Also known as a star-mesh transformation)

The theorem states that at any given frequency a star network can be interchanged with a delta network (as in Fig.5.7) provided that certain relationships between the elements of the two networks are maintained. These are developed below:

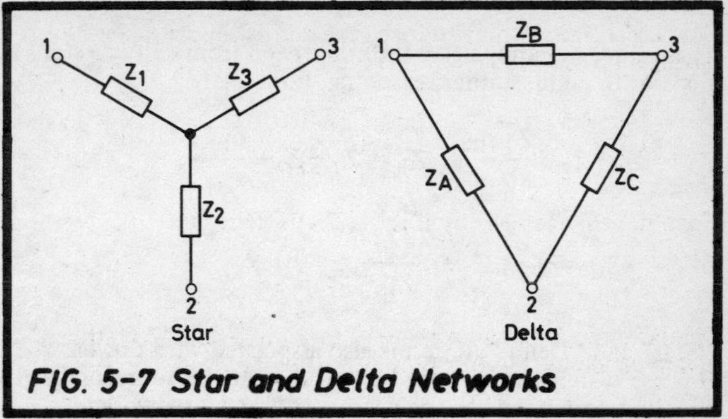

FIG. 5-7 Star and Delta Networks

Considering the equivalence of the impedance looking into terminals 1 and 2 of both networks,

$$Z_1 + Z_2 = \frac{Z_A(Z_B + Z_C)}{Z_A + Z_B + Z_C} \quad ...(1)$$ (2 resistors in parallel, Sect.3.2.3)

Terminals 1 and 3:

$$Z_1 + Z_3 = \frac{Z_B(Z_A + Z_C)}{Z_A + Z_B + Z_C} \quad ...(2)$$

Terminals 2 and 3:

$$Z_2 + Z_3 = \frac{Z_C(Z_A + Z_B)}{Z_A + Z_B + Z_C} \quad ...(3)$$

from which, by adding equations (1) and (2) and subtracting (3)

$$Z_1 = \frac{Z_A Z_B}{Z_A + Z_B + Z_C} \quad \text{and similarly} \quad Z_2 = \frac{Z_A Z_C}{Z_A + Z_B + Z_C}$$

$$\text{and} \quad Z_3 = \frac{Z_B Z_C}{Z_A + Z_B + Z_C}$$

These three equations give the conversion from delta to star. For star to delta, similar reasoning shows that:

$$Z_A = Z_1 + Z_2 + \frac{Z_1 Z_2}{Z_3}, \quad Z_B = Z_1 + Z_3 + \frac{Z_1 Z_3}{Z_2},$$

$$Z_C = Z_2 + Z_3 + \frac{Z_2 Z_3}{Z_1}$$

"Star" and "Delta" are terms also associated with power 3-phase systems and it may be that they are better recognized in telecommunications engineering as T and π-networks respectively. Their equivalence is easily seen by comparing Fig.5.7 with Fig.5.8.

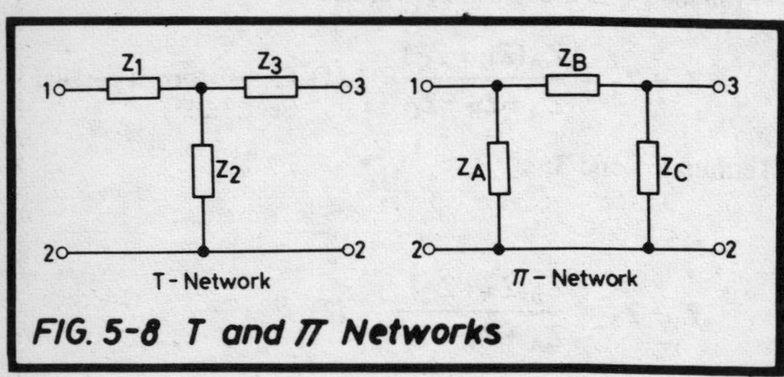

FIG. 5-8 T and π Networks

Examples of conversion are not given here because the principle is employed in Sect.5.3.4 to change from one network to the other and an example is given.

5.2 WAVEFORM ANALYSIS

Analysis of highly complex waveforms is not of much practical use in the amateur workshop because the equipment required for separating out the various harmonics is expensive.

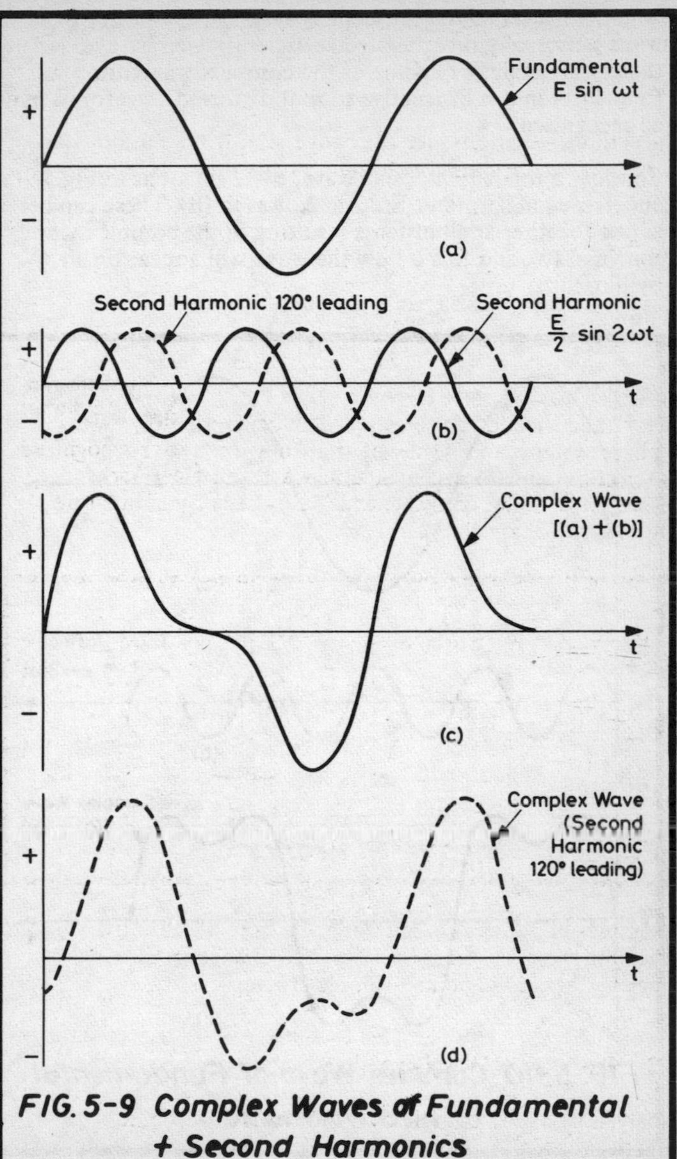

FIG. 5-9 Complex Waves of Fundamental + Second Harmonics

When, however, an oscilloscope is available the purity of a sine wave can usually be estimated sufficiently well by eye. Nevertheless, an analysis of some of the complex waveforms more frequently met is instructive so that distorted waveforms may be recognised.

Consider a fundamental sine wave, $e = E \sin \omega t$ as in Fig.5.9(a) and its second harmonic $E/2 \sin 2\omega t$ as in (b). These can be added together at all instants resulting in the complex wave shown at (c), and this is how the wave will appear on an oscilloscope.

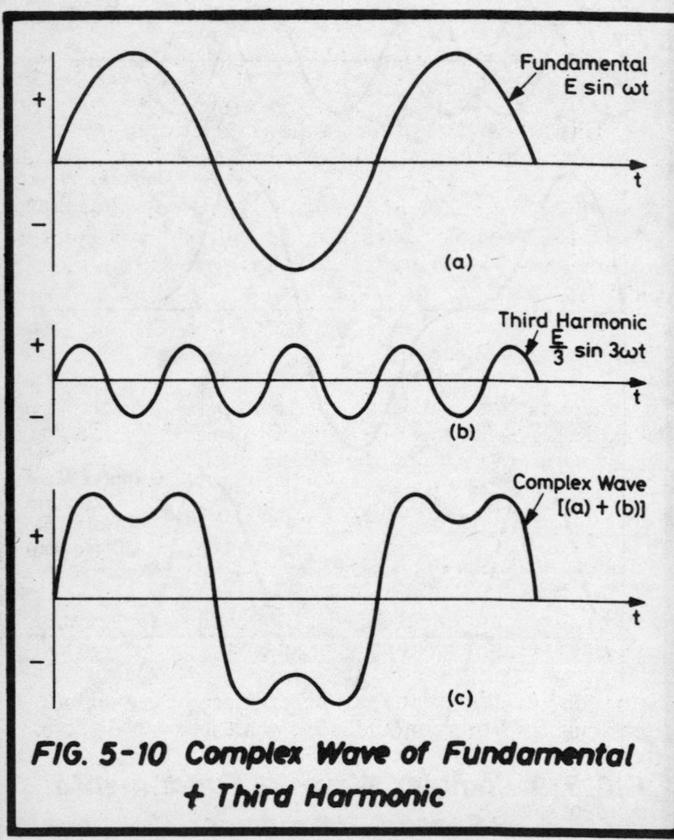

FIG. 5-10 Complex Wave of Fundamental + Third Harmonic

Also shown dotted on (b) is the same harmonic but leading by 120° (on its own angular scale, not on that of the fundamental). The addition of this wave with the fundamental to form a complex wave is shown at (d).

Fig.5.10 repeats the process for a fundamental wave with its third harmonic in-phase.

These are the simplest forms of complex waves, consisting as they do, of a fundamental and only one harmonic but they help to demonstrate the dependence of a complex wave on

(i) the number of odd and/or even harmonics present

(ii) the relative amplitudes of fundamental and harmonics

(iii) the phases of the harmonics relative to the fundamental and to themselves.

Generally, the presence of even harmonics results in an asymmetrical complex wave, whereas with odd harmonics, a symmetrical wave is produced. This is evident from Figs.5.9 and 5.10.

5.2.1 Fourier's Theorem

The theorem states that any continuous periodic function can be expressed as the sum of a number of sine waves of differing frequency and amplitude, i.e. in voltage terms

$$e = c + E_1 \sin(\omega t + \phi_1) + E_2 \sin(2\omega t + \phi_2) + E_3 \sin(3\omega t + \phi_3) + \ldots$$

where c is a constant and E_1, E_2, E_3 etc. are the maximum voltage values of the various components.

Some simplification in the mathematics can be gained by producing constants containing the phase angles ϕ_1, ϕ_2 etc. thus:

Since, for example,

$$E_1 \sin(\omega t + \phi_1) = E_1 \sin \omega t \cos \phi_1 + E_1 \cos \omega t \sin \phi_1$$

if $E_1 \cos \phi_1$ is represented by a_1
and $E_1 \sin \phi_1$ is represented by b_1 etc.

$$\text{Then } e = c + a_1 \sin \omega t + a_2 \sin 2\omega t + a_3 \sin 3\omega t + \ldots\ldots + b_1 \cos \omega t + b_2 \cos 2\omega t + b_3 \cos 3\omega t + \ldots\ldots$$

This is the general expression from which the components of complex waveforms may be deduced. Some in common use are as follows:

(1) *Square Waveform*

This is illustrated in Fig.5.11. It has a maximum value of E and, as shown, the frequency of the fundamental is $\omega/2\pi$ Hz.

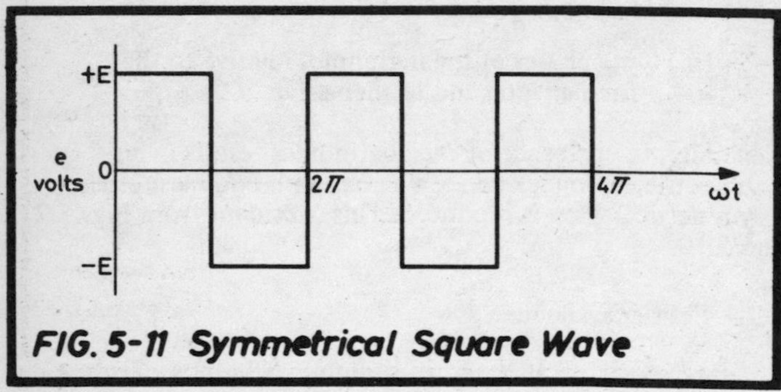

FIG. 5-11 Symmetrical Square Wave

Determination of the values of the coefficients in the above expression gives:

c (which is in fact, the mean value of the curve over one cycle) = 0

$$a_1 = \frac{4E}{\pi} \quad a_3 = \frac{4E}{3\pi} \quad a_5 = \frac{4E}{5\pi} \ldots\ldots$$

$$a_2 = 0 \quad a_4 = 0 \quad a_6 = 0 \ldots\ldots$$

all b terms = 0.

$$\text{Therefore } e = \frac{4E}{\pi}(\sin \omega t + \tfrac{1}{3} \sin 3\omega t + \tfrac{1}{5} \sin 5\omega t + \ldots\ldots) \text{ volts.}$$

This is shown graphically in Fig.5.12 and from the dotted curve in the first half-cycle of the top graph, the squaring of the sine curve by the addition of the 3rd and 5th harmonics only, is already becoming evident.

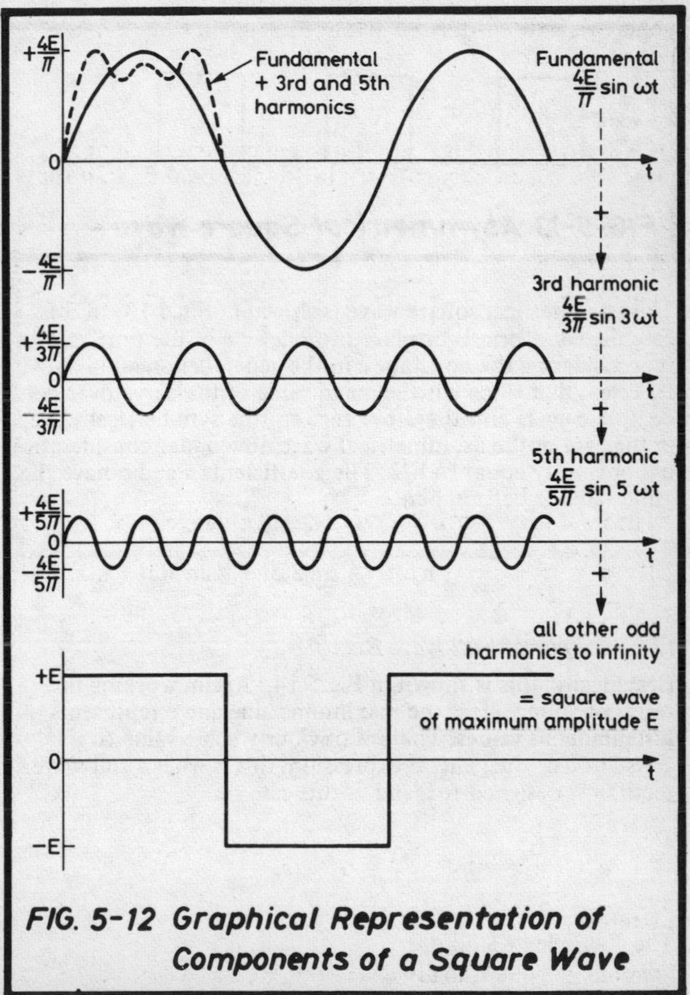

FIG. 5-12 Graphical Representation of Components of a Square Wave

Thus the square wave consists of a fundamental wave with an infinite series of odd harmonics progressively decreasing in peak value. This conception is important because overloading in amplifiers and systems usually results in "squaring" a waveform (e.g. cutting off the top of a sine wave), thus producing odd harmonics.

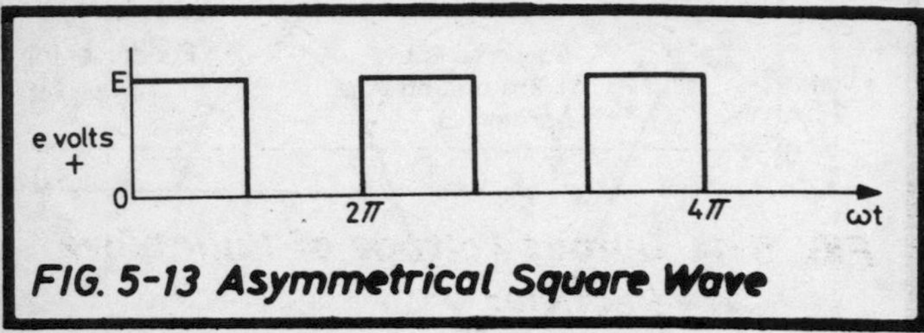

FIG. 5-13 Asymmetrical Square Wave

The asymmetrical square wave is shown in Fig.5.13. In this case the waveform is displaced to one side of the time axis and on considering the constant c in the general expression above, it is clear that since c is the mean value of the curve over one complete cycle and therefore zero for the symmetrical wave, in the case of the asymmetrical wave now under consideration, c is obviously equal to E/2. The coefficients a and b have the same value as before, hence

$$e = \frac{E}{2} + \frac{2E}{\pi}\left(\sin \omega t + \tfrac{1}{3}\sin 3\omega t + \tfrac{1}{5}\sin 5\omega t + \ldots\ldots\right) \text{ volts.}$$

(2) *Output of Full-Wave Rectifier*

Graphically, this is shown in Fig.5.14. Again, working in terms of voltage, E is the maximum value and e represents instantaneous values. There is obviously some value to the constant c of the general expression, this is what a full-wave rectifier is designed for, and in this case

$$c = \frac{2E}{\pi}$$

The coefficients a all = 0
" " $b_1, b_3, b_5 \ldots\ldots = 0$
" " $b_2, b_4, b_6 \ldots\ldots$ are given by the overall expressio[n]

$$-\frac{4E}{\pi} \cdot \frac{\cos n\omega t}{(n+1)(n-1)}$$

where $n = 2, 4, 6 \ldots\ldots$

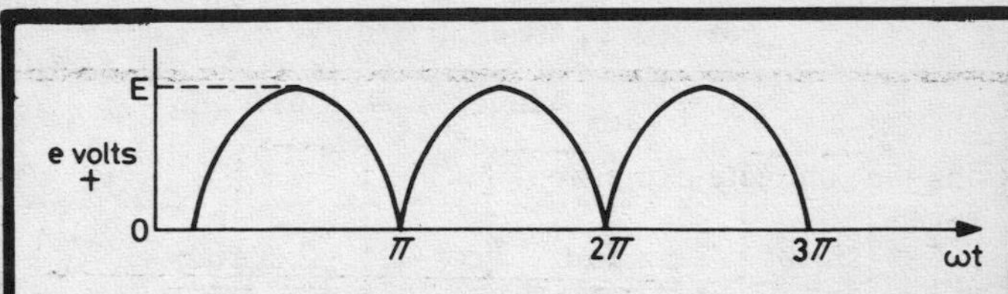

FIG. 5-14 Output Voltage of Full-Wave Rectifier

Hence the complete expression for the components of Fig.5.14 is

$$e = \frac{2E}{\pi} - \frac{4E}{3\pi}\cos 2\omega t - \frac{4E}{15\pi}\cos 4\omega t - \frac{4E}{35\pi}\cos 6\omega t - \ldots\ldots$$

i.e. $$e = \frac{2E}{\pi} - \frac{4E}{\pi}\left\{\frac{\cos 2\omega t}{3} + \frac{\cos 4\omega t}{15} + \frac{\cos 6\omega t}{35} + \ldots\ldots\right\} \text{volts.}$$

(3) *Output of Half-Wave Rectifier*

This is shown graphically in Fig.5.15.

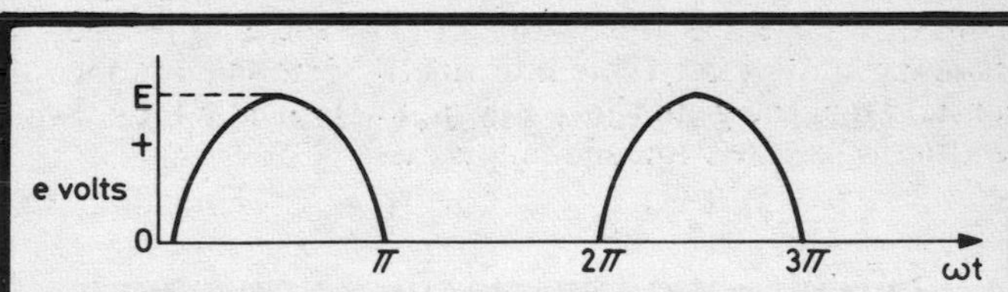

FIG. 5-15 Output voltage of Half-Wave Rectifier

As in the full-wave case the constant c has a value, but now reduced to half, i.e. $c = E/\pi$.

Also $a_1 = \dfrac{E}{2}$ $\quad a_2, a_3, a_4 \ldots\ldots = 0$

$b_1, b_3, b_5 \ldots\ldots = 0$

b_2, b_4, b_6 are expressed by $-\dfrac{2E}{\pi} \cdot \dfrac{\cos n\omega t}{(n+1)(n-}$

where $n = 2, 4, 6 \ldots\ldots$

giving the complete expression:

$$e = \frac{E}{\pi} + E\left\{\frac{\sin\omega t}{2} - \frac{2\cos 2\omega t}{3\pi} - \frac{2\cos 4\omega t}{15\pi} - \frac{2\cos 6\omega t}{35\pi} - \ldots\ldots\right\} \text{vol}$$

(4) *Saw-Tooth Waveform*

By following the above principles, see Fig.5.16

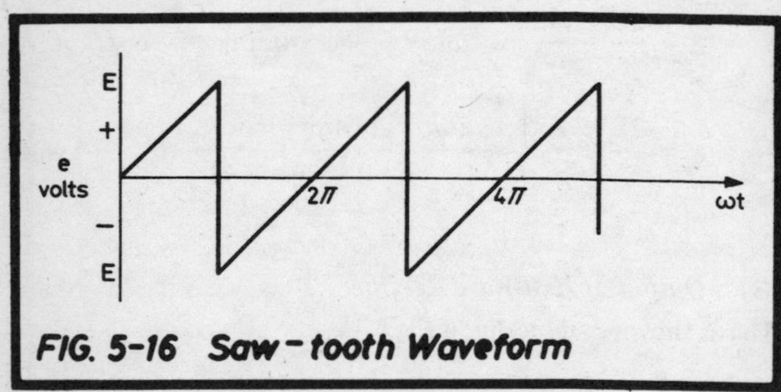

FIG. 5-16 Saw-tooth Waveform

$$e = \frac{2E}{\pi}\left\{\sin\omega t - \tfrac{1}{2}\sin 2\omega t + \tfrac{1}{3}\sin 3\omega t - \tfrac{1}{4}\sin 4\omega t + \ldots\ldots\right\} \text{volts.}$$

5.3 ATTENUATING AND MATCHING NETWORKS

These exist in several forms, L, T, π etc., symmetrical, asymmetrical, balanced and unbalanced, the latter terms are first defined.

Symmetrical networks are used to provide attenuation between equal impedances, accordingly the pad itself has equal resistances on both sides.

Asymmetrical networks provide attenuation between unequal impedances and the pad resistances on the two sides are unequal.

Balanced – the series arm is divided equally between the two legs of the circuit.

Unbalanced – the series arm of the pad is placed in one leg of the circuit only.

These features are demonstrated by the various configurations of, for example, a T-type pad in Fig.5.17.

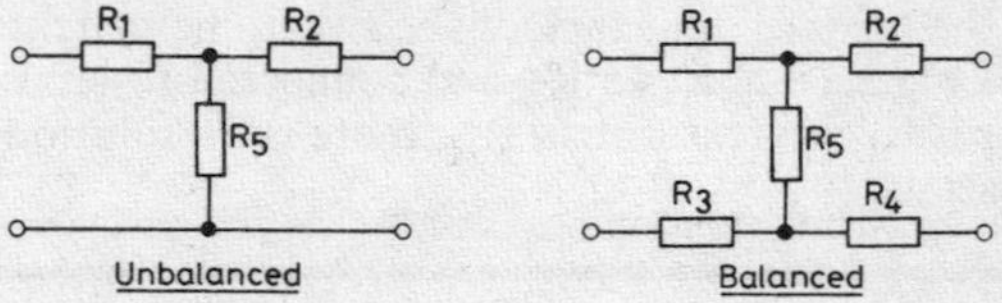

R1 R2 R5 Unbalanced

R1 R2 R5 R3 R4 Balanced

Symmetrical, Unbalanced $R_1 = R_2$; Balanced $R_1 = R_2 = R_3 = R_4$
Asymmetrical, Unbalanced $R_1 \neq R_2$; Balanced $R_1 = R_3$, $R_2 = R_4$
$R_1 \neq R_2$, $R_3 \neq R_4$

FIG. 5-17 Configurations of T-type Networks

The pads usually contain resistors only so that the attenuation is independent of frequency.

5.3.1 L-type Network

The L-type is the simplest, minimum attenuation network used for matching two different impedances e.g. R_H to R_L (high to low) as in Fig.5.18.

To match R_H, terminals 1 and 2 must present an impedance R_H hence

$$R_1 + \frac{R_2 R_L}{R_2 + R_L} = R_H \qquad (1)$$

and at terminals 3 and 4

$$\frac{R_2(R_H + R_1)}{R_2 + R_1 + R_H} = R_L \qquad (2)$$

Adding equations (1) and (2) gives $R_1 R_2 = R_H R_L$

$$\therefore\ R_1 = \frac{R_H R_L}{R_2}$$

and by further substitution:

$$R_1 = \sqrt{R_H(R_H - R_L)} \qquad R_2 = \sqrt{\frac{R_H R_L^2}{R_H - R_L}}$$

Clearly, the higher impedance of the two to be matched by the circuit of Fig.5.18 must be R_H and connected to terminals 1 and 2 for if R_L is greater than R_H, there is no solution to the two equations.

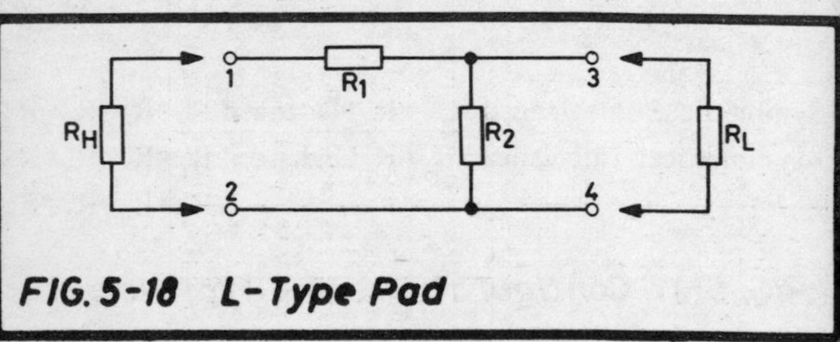

FIG. 5-18 L-Type Pad

As a guide, Table 5.1 gives a range of L-type pad values for values of R_H/R_L up to 100:1. The whole table can be multiplied by multiples of 10 as required.

Attenuation

Calculation of the attenuation (or "insertion loss") of the pad is simplified by using the principle of the division of current in a 2-resistor parallel combination (Sect.3.2.4). Consider the circuit of Fig.5.19 in which R_1 and R_2 form an L-type pad matching R_H to R_L.

R_L Ω \ R_H Ω	1		2		3		4		5		6		7		8		9		10	
	R_1	R_2	R_1	R_2	R_1	R_2	R_1	R_2	R_1	R_2	R_1	R_2	R_1	R_2	R_1	R_2	R_1	R_2	R_1	R_2
1	0	∞	1.414	1.414	2.450	1.225	3.464	1.155	4.472	1.118	5.477	1.095	6.481	1.080	7.483	1.069	8.485	1.061	9.487	1.054
2			0	∞	1.732	3.464	2.828	2.828	3.873	2.582	4.899	2.450	5.916	2.366	6.928	2.309	7.937	2.268	8.944	2.236
3					0	∞	2.000	6.000	3.162	4.743	4.243	4.243	5.292	3.969	6.325	3.795	7.349	3.674	8.367	3.586
4							0	∞	2.236	8.944	3.464	6.928	4.583	6.110	5.657	5.657	6.708	5.367	7.746	5.164
5									0	∞	2.450	12.25	3.742	9.354	4.899	8.165	6.000	7.500	7.071	7.071
6											0	∞	2.646	15.87	4.000	12.00	5.196	10.39	6.325	9.487
7													0	∞	2.828	19.80	4.243	14.85	5.477	12.78
8															0	∞	3.000	24.00	4.479	17.89
9																	0	∞	3.162	28.46
10																			0	∞

R_L Ω \ R_H Ω	20		30		40		50		60		70		80		90		100	
	R_1	R_2	R_1	R_2	R_1	R_2	R_1	R_2	R_1	R_2	R_1	R_2	R_1	R_2	R_1	R_2	R_1	R_2
1	19.49	1.026	29.50	1.017	39.50	1.013	49.50	1.010	59.50	1.008	69.50	1.007	79.50	1.006	89.50	1.006	99.50	1.005
2	18.97	2.108	28.98	2.070	38.99	2.052	48.99	2.041	58.99	2.034	68.99	2.029	78.99	2.026	88.99	2.023	99.00	2.020
3	18.44	3.254	28.46	3.162	38.47	3.119	48.48	3.094	58.48	3.078	68.48	3.066	78.49	3.058	88.49	3.051	98.49	3.046
4	17.89	4.472	27.93	4.297	37.95	4.216	47.96	4.170	57.97	4.140	67.97	4.119	77.97	4.104	87.98	4.092	97.98	4.083
5	17.32	5.774	27.39	5.477	37.42	5.345	47.43	5.271	57.45	5.222	67.45	5.189	77.46	5.164	87.46	5.145	97.47	5.130
6	16.73	7.171	26.83	6.708	36.88	6.508	46.90	6.396	56.92	6.325	66.93	6.275	76.94	6.239	86.95	6.211	96.95	6.189
7	16.12	8.682	26.27	7.995	36.33	7.707	46.37	7.548	56.39	7.448	66.41	7.379	76.42	7.328	86.43	7.289	96.44	7.259
8	15.49	10.33	25.69	9.342	35.78	8.944	45.83	8.729	55.86	8.593	65.88	8.501	75.90	8.433	85.91	8.381	95.92	8.341
9	14.83	12.14	25.10	10.76	35.21	10.22	45.28	9.939	55.32	9.762	65.35	9.641	75.37	9.553	85.38	9.487	95.39	9.435
10	14.14	14.14	24.50	12.25	34.64	11.55	44.72	11.18	54.77	10.95	64.81	10.80	74.83	10.69	84.85	10.61	94.87	10.54

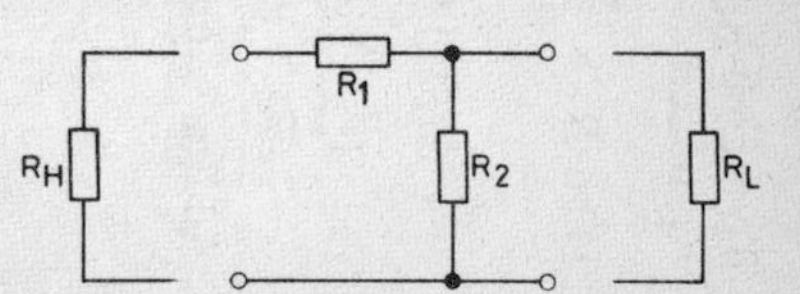

Table 5.1 L-Type Pad Values (ohms) for Matching R_H to R_L

Let I_H be the current in R_H and I_L the current in R_L.

$$\text{Then } I_L = I_H \cdot \frac{R_2}{R_L + R_2}$$

$$\therefore \frac{I_H}{I_L} = \frac{R_L + R_2}{R_2}$$

and the attenuation in decibels $= 10 \log_{10} \dfrac{I_H^2 R_H}{I_L^2 R_L}$ (Sect.6.2.6)

$$= 10 \log_{10} \left(\frac{R_L + R_2}{R_2}\right)^2 \cdot \frac{R_H}{R_L}$$

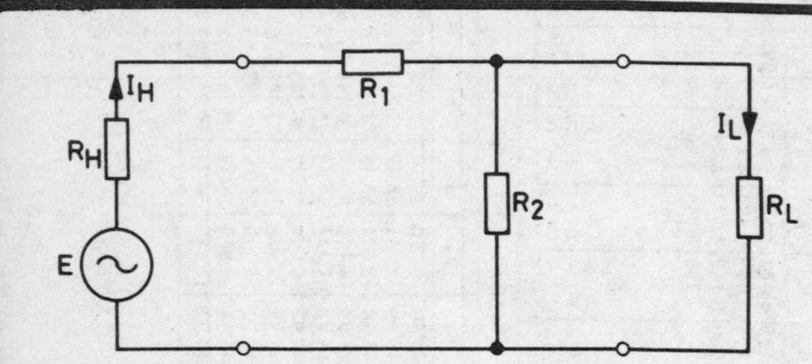

FIG. 5-19 Currents in Terminations of L-Type Pad

Th. attenuations of the pads shown in Table 5.1 vary from about 2.5 dB for the smallest matching ratio (R_H/R_L) to over 30 dB for the highest. Table 5.2 shows this expressed to the nearest 0.5 dB as a function of matching ratio. The approximate values are useful to check calculations or for example, when a choice is to be made between pad and transformer matching.

[The attenuation can also be obtained from Table 5.5 in the next section by substituting R_H/R_L for $(n_2/n_1)^2$ in the heading of the last column, e.g. for a matching ratio of 3.023, the L-type pad attenuation is 10 dB. This arises from the fact that when $R_H/R_L = (n_2/n_1)^2$, a T-pad is reduced to an L-type, hence the attenuation quoted in the table is appropriate.]

Matching Ratio R_H/R_L	Pad Attenuation dB	Matching Ratio R_H/R_L	Pad Attenuation dB
1.1	2.5	7.0	14.0
1.2	3.5	8.0	15.0
1.3	4.5	9.0	15.5
1.4	5.0	10	16.0
1.5	5.5	12	16.5
1.6	6.0	14	17.5
1.7	6.5	16	18.0
1.8	7.0	18	18.5
1.9	7.5	20	19.0
2.0	7.5	25	20.0
2.2	8.0	30	20.5
2.4	9.0	40	22
2.6	9.0	50	23
2.8	9.5	60	24
3.0	10.0	80	25
3.5	11.0	100	26
4.0	11.5	150	28
5.0	12.5	200	29
6.0	13.5	250	30

Table 5.2 **Approximate L-Type Pad Attenuations for Matching Ratios R_H/R_L**

EXAMPLE:
A coaxial line of 70 ohms must be correctly terminated with 70 ohms to minimize reflexions. The input impedance of the amplifier to which the line is to be connected is 400 ohms. Design a matching pad having minimum attenuation. What is the attenuation?

For minimum attenuation, the L-type pad is used, see Fig.5.20.

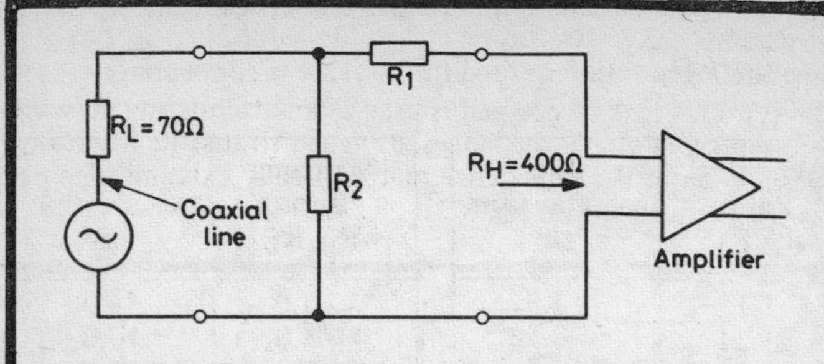

FIG. 5-20 Coaxial Line Matched to Amplifier

From Table 5.1 for $R_H = 40$ ohms, $R_L = 7$ ohms,

$$R_1 = 36.33 \text{ ohms}, \quad R_2 = 7.707 \text{ ohms}$$

Therefore, for $R_H = 400$ ohms, $R_L = 70$ ohms

$$R_1 = 363.3 \text{ ohms} \quad R_2 = 77.1 \text{ ohms}$$

The nearest preferred value resistors e.g. 360 ohms and 75 ohms are likely to produce sufficiently accurate matching, but if greater precision were to be sought, then R_2 can be made up from two resistors, 62 and 15 ohms (both are preferred values) in series to give 77 ohms.

$$\text{Attenuation} = 10 \log_{10} \left(\frac{R_L + R_2}{R_2}\right)^2 \cdot \frac{R_H}{R_L}$$

$$= 10 \log_{10} \left(\frac{70 + 77.1}{77.1}\right)^2 \cdot \frac{400}{70} = 10 \log_{10} 20.8$$

$$= 13.18 \text{ dB.}$$

This can be checked from Table 5.2 which shows that for $R_H/R_L = 400/70 = 5.7$, the approximate pad attenuation is just over 13 dB.

5.3.2 T-type Network

Symmetrical

The symmetrical T-type pad is used as an attenuating network between two equal impedances, designed so that its insertion does not upset the impedance match already existing. The pad has the form shown in Fig.5.21.

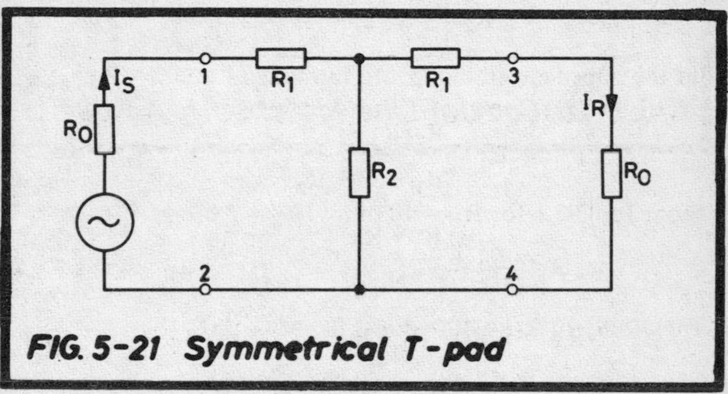

FIG. 5-21 Symmetrical T-pad

From the previous section it is seen that the L-pad, being asymmetrical, is incapable of maintaining the matching between two equal impedances, it can only function between two unequal ones. Hence the symmetrical T-pad which in effect is similar to the L-type but with the series arm equally divided on both sides of the shunt arm. Because it can maintain the matching, the symmetrical T-pad can be designed for attenuations other than the minimum.

The starting point in the design of a symmetrical T-pad is the attenuation required, but in analysis of the network, voltages, currents and resistances only are considered (attempting to work directly in decibels is an unnecessary complication involving logarithmic equations) so it is most convenient to convert the attenuation into the current ratio first. If this is designated by N and the pad attenuation required is α dB,

then since $\alpha = 20 \log_{10} \dfrac{I_S}{I_R} = 20 \log_{10} N$

$$\therefore N = \text{antilog} \frac{\alpha}{20}$$

Considering Fig.5.21 and using the principle of the division of current in a 2-resistor parallel combination (Sect.3.2.4)

$$I_R = I_S \cdot \frac{R_2}{R_1 + R_2 + R_o}$$

$$\therefore \frac{I_S}{I_R} = \frac{R_1 + R_2 + R_o}{R_2} = N$$

But the impedance looking into terminals 1 and 2 = R_o

$$\text{i.e.} \quad R_o = R_1 + \frac{R_2(R_1 + R_o)}{R_1 + R_2 + R_o}$$

$$= R_1 + \frac{R_1 + R_o}{N}$$

$$\therefore \quad NR_o = NR_1 + R_1 + R_o$$

$$\therefore \quad R_1 = R_o \cdot \frac{(N-1)}{(N+1)}$$

and by further substitution:

$$R_2 = R_o\left(\frac{2N}{N^2 - 1}\right)$$

Now let $\dfrac{N-1}{N+1} = n_1$ and $\dfrac{2N}{N^2-1} = n_2$

to make Table 5.3 possible. This is constructed on the basis of values for n_1 and n_2 being given for a range of pad attenuations, from which design is greatly simplified, i.e. by multiplying the terminating impedance by n_1 and n_2 in turn, the values of R_1 and R_2 are given directly.

EXAMPLE:
A 3-position variable attenuator is required to work between 600 ohm impedances, having values of 3, 6 and 10 dB. Design the circuit using T-type networks.

α_1 dB	T-type Pad		π-type Pad	
	n_1	n_2	n_1	n_2
1	.0575	8.667	17.39	.1154
2	.1147	4.304	8.722	.2324
3	.1708	2.842	5.854	.3519
4	.2263	2.096	4.419	.4770
5	.2800	1.645	3.571	.6078
6	.3322	1.339	3.010	.7469
7	.3825	1.116	2.614	.8962
8	.4305	.9461	2.323	1.057
9	.4762	.8120	2.100	1.232
10	.5195	.7028	1.925	1.423
11	.5602	.6124	1.785	1.633
12	.5985	.5362	1.671	1.865
13	.6342	.4714	1.577	2.122
14	.6673	.4156	1.499	2.406
15	.6980	.3673	1.433	2.723
16	.7264	.3251	1.377	3.076

α_2 dB	T-type Pad		π-type Pad	
	n_1	n_2	n_1	n_2
17	.7524	.2883	1.329	3.469
18	.7764	.2559	1.288	3.909
19	.7982	.2273	1.253	4.400
20	.8182	.2020	1.222	4.950
22	.8528	.1599	1.173	6.255
24	.8813	.1267	1.135	7.893
26	.9045	.1005	1.106	9.951
28	.9234	.0797	1.083	12.54
30	.9387	.0633	1.065	15.80
35	.9651	.0356	1.036	28.11
40	.9802	.0200	1.020	50.00
45	.9888	.0112	1.011	88.91
50	.9937	.0063	1.006	158.1
55	.9964	.0036	1.004	281.2
60	.9980	.0020	1.002	500.0

Design:

T-type R_o ← R_1 R_2 R_1 → R_o

$R_1 = R_o \times n_1$

$R_2 = R_o \times n_2$

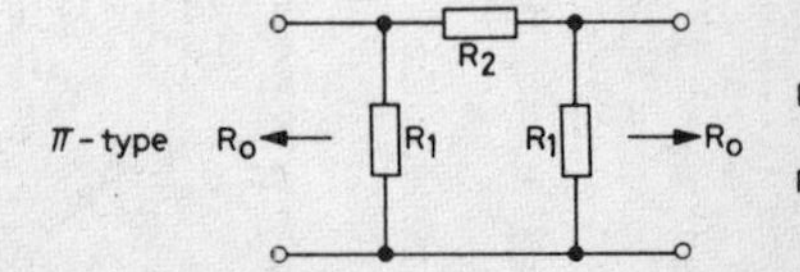

$R_1 = R_o \times n_1$

$R_2 = R_o \times n_2$

Table 5.3 **Values of Multiplying Factors, n for Symmetrical T and π-type Pads**

Calculation of Resistor Values. From Table 5.3:

3 dB pad $n_1 = .1708 \therefore R_1 = 600 \times .1708 = 102.5$ ohms
$n_2 = 2.842 \therefore R_2 = 600 \times 2.842 = 1{,}705$ ohms

6 dB pad $n_1 = .3322$ $\therefore R_1 = 199.3$ ohms
$n_2 = 1.339$ $R_2 = 803.4$ ohms

10 dB pad $n_1 = .5195$ $R_1 = 311.7$ ohms
$n_2 = .7028$ $R_2 = 421.7$ ohms

The variable attenuator is therefore as shown in Fig.5.22.

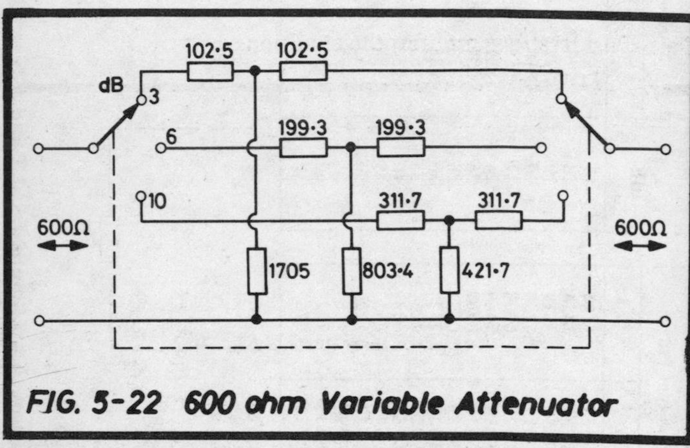

FIG. 5-22 600 ohm Variable Attenuator

Asymmetrical

This type of T-pad is capable of matching together two unequal impedances R_H and R_L and at the same time inserting a specified attenuation between them, see Fig.5.23. As shown later, there is a certain minimum design loss which is dependent on the ratio of the termination impedances, i.e. R_H/R_L.

R_H is the higher of the two terminating impedances and R_L, the lower.

The formulae for calculation of the pad resistances are naturally more complicated than for the symmetrical case because there are more parameters to be handled. For example, the attenuation cannot be resolved as a current ratio because the two impedances differ, nevertheless, for the asymmetrical case if

$$N = \sqrt{\frac{P_H}{P_L}}$$ where P_H and P_L are the two powers,

then as before:

$$N = \text{antilog}\,\frac{\alpha}{20}$$ where α dB is the attenuation of the pad.

This considerably simplifies the equations.

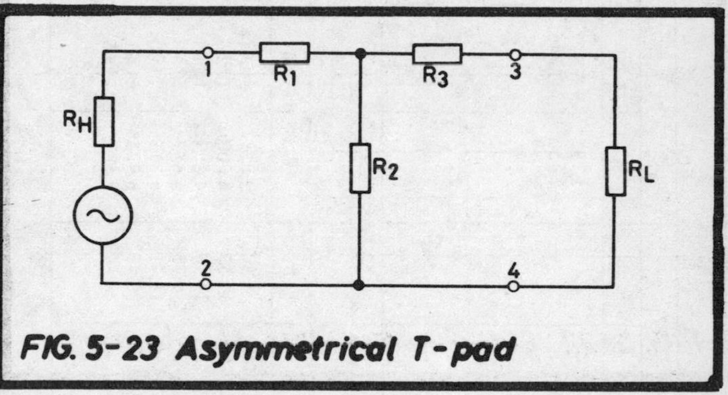

FIG. 5-23 Asymmetrical T-pad

By similar reasoning as in the symmetrical case:

$$R_2 = \sqrt{R_H R_L}\left(\frac{2N}{N^2-1}\right)$$

$$R_1 = R_H\left(\frac{N^2+1}{N^2-1}\right) - R_2$$

$$R_3 = R_L\left(\frac{N^2+1}{N^2-1}\right) - R_2$$

$R_L \downarrow$ Ω \ $R_H \rightarrow$ Ω	1		2		3		4		5	
	A	B	A	B	A	B	A	B	A	B
1	1.000	3.162	1.414	4.472	1.732	5.477	2.000	6.325	2.236	7.071
2					2.450	7.746	2.828	8.944	3.162	10.00
3							3.464	10.95	3.873	12.25
4									4.472	14.14
5										
6										
7										
8										
9										

$R_L \downarrow$ Ω \ $R_H \rightarrow$ Ω	6		7		8		9		10	
	A	B	A	B	A	B	A	B	A	B
1	2.450	7.746	2.646	8.366	2.828	8.944	3.000	9.487	3.162	10.00
2	3.464	10.95	3.742	11.83	4.000	12.65	4.243	13.42	4.472	14.14
3	4.243	13.42	4.583	14.49	4.899	15.49	5.196	16.43	5.477	17.32
4	4.899	15.49	5.292	16.73	5.657	17.89	6.000	18.97	6.325	20.00
5	5.477	17.32	5.916	18.71	6.325	20.00	6.708	21.21	7.071	22.36
6			6.481	20.49	6.928	21.91	7.349	23.24	7.746	24.50
7					7.483	23.66	7.937	25.10	8.367	26.46
8							8.485	26.83	8.944	28.28
9									9.487	30.00

MULTIPLY COLUMN A OR B FIGURES AS SHOWN:

$R_H \rightarrow$ Ω / $R_L \downarrow$ Ω	1-10		10-100		100-1000		1000-10,000		10^4-10^5	
	A	B	A	B	A	B	A	B	A	B
1-10	NIL			NIL	x 10			x 10	x 100	
10-100			x 10			x 10	x 100			x 100
100-1000					x 100			x 100	x 1000	
1000-10,000							x 1000			x 1000

NIL – indicates column to be used with no correction.

R_H = Terminating Impedance (higher value)
R_L = Terminating Impedance (lower value)

Table 5.4 **Attenuator Terminating Impedances – Values of $\sqrt{R_H R_L}$**

Tables can therefore be constructed to facilitate design on the basis that

$$\frac{2N}{(N^2-1)} = n_1 \quad \text{and} \quad \frac{N^2+1}{N^2-1} = n_2$$

Then $R_2 = \sqrt{R_H R_L} \times n_1$

$R_1 = n_2 R_H \quad - R_2$

$R_3 = n_2 R_L \quad - R_2$

For calculation of R_2, Table 5.4 may be of use to quickly determine $\sqrt{R_H R_L}$ for example, if R_H is 900 ohms and R_L 50 ohms, then at $R_H = 9$, $R_L = 5$ in the first section of the table, the figures

A 6.708 B 21.21 are given while the second section

of the table shows that for R_H in 100-1000 ohm range and R_L in the 10-100 ohm range, the B figure should be used and multiplied by 10, thus

$$\sqrt{900 \times 50} = 212.1 .$$

Table 5.5 is subsequently used to give the value of n_1 and this is multiplied by the figure for $\sqrt{R_H R_L}$ to calculate R_2. With the value of R_2 determined, R_1 and R_3 follow from the figure for n_2, again from Table 5.5.

Minimum Pad Loss

As an asymmetrical T-pad is progressively designed for lower pad attenuations, R_3 gets smaller and the limit is reached when $R_3 = 0$ because for an attenuation lower than this, R_3 has a negative value which cannot be achieved in a practical circuit. In fact, at $R_3 = 0$ the T-pad has been transformed into an L-pad which, as previously shown has a fixed attenuation for any given ratio R_H/R_L. This indicates that the asymmetrical T-pad has a minimum loss, so that design of a pad matching R_H to R_L for a lower attenuation than this minimum is not possible. This value can be determined by considering the condition $R_3 = 0$ which occurs when $n_2 R_L = R_2$ or

$$n_2 R_L = n_1 \sqrt{R_H R_L}$$

$$\therefore \frac{R_H}{R_L} = \left(\frac{n_2}{n_1}\right)^2$$

Table 5.5 includes the values of $(n_2/n_1)^2$ so that comparison may be made with R_H/R_L to ensure that

$$\frac{R_H}{R_L} < \left(\frac{n_2}{n_1}\right)^2$$

for any particular magnitude of attenuation required.

α, dB	n_1	n_2	$\left(\frac{n_2}{n_1}\right)^2$	α, dB	n_1	n_2	$\left(\frac{n_2}{n_1}\right)^2$
1	8.667	8.725	1.013	17	.2883	1.041	13.04
2	4.304	4.418	1.054	18	.2559	1.032	16.26
3	2.842	3.013	1.124	19	.2273	1.026	20.38
4	2.096	2.323	1.228	20	.2020	1.020	25.50
5	1.645	1.925	1.369	22	.1599	1.013	40.14
6	1.339	1.671	1.557	24	.1267	1.008	63.30
7	1.116	1.498	1.802	26	.1005	1.005	100.0
8	.9461	1.377	2.118	28	.0797	1.003	158.4
9	.8120	1.288	2.516	30	.0633	1.002	250
10	.7028	1.222	3.023	35	.0356	1.0006	790
11	.6124	1.173	3.669	40	.0200	1.0002	2501
12	.5362	1.135	4.481	45	.0112	1.00006	
13	.4714	1.106	5.505	50	.0063	1.00002	
14	.4156	1.083	6.791	55	.0036	$\simeq$ 1.0	
15	.3673	1.065	8.407	60	.0020	$\simeq$ 1.0	
16	.3251	1.052	10.47				

Design:

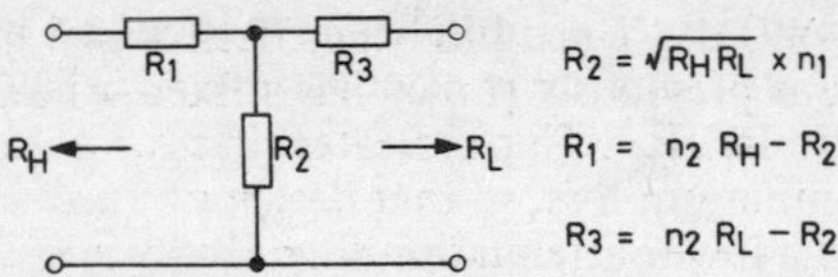

Table 5.5 **Values of Multiplying Factors, n, for Asymmetrical T-type Pads**

EXAMPLE:
A 75 ohm line feeds a loudspeaker of 40 ohms. A control on the loudspeaker is required to give full volume with optional 10 and 20 dB reductions. Design a practical T-type attenuator to match the line to the loudspeaker and to give the required volume reductions.

First check for minimum pad loss:

$$\frac{R_H}{R_L} = \frac{75}{40} = 1.875$$

From Table 5.5, for 10 dB attenuation R_H/R_L must be equal to, or less than 3.023. This is so, therefore a pad can be designed for 10 dB loss.

Design 10 dB pad – from Table 5.5 $n_1 = .7028$, $n_2 = 1.222$

$$R_2 = \sqrt{75 \times 40} \times .7028 = 38.49 \text{ ohms}$$
$$R_1 = (1.222 \times 75) - 38.49 = 53.16 \text{ ''}$$
$$R_3 = (1.222 \times 40) - 38.49 = 10.39 \text{ ''}$$

Design 20 dB pad – from Table 5.5 $n_1 = .2020$, $n_2 = 1.020$

$$R_2 = \sqrt{75 \times 40} \times .2020 = 11.06 \text{ ohms}$$
$$R_1 = (1.020 \times 75) - 11.06 = 65.44 \text{ ''}$$
$$R_3 = (1.020 \times 40) - 11.06 = 29.74 \text{ ''}$$

A suitable circuit would therefore be as follows (Fig.5.24).

The nearest preferred-value resistor has been used in each case. Considering that the output is sound from a loudspeaker and the ear can only detect a 1 dB change in loudness with great difficulty, the discrepancy in attenuation is negligible.

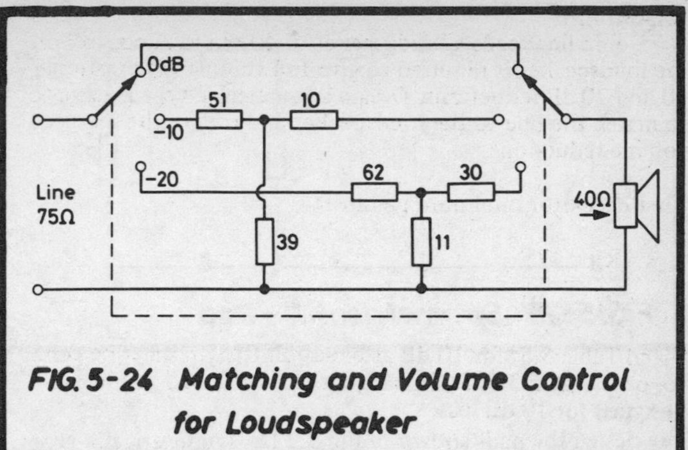

FIG. 5-24 Matching and Volume Control for Loudspeaker

5.3.3 π-type Network

Symmetrical

Whereas the T-pad consists of two series and one shunt arm, the π-pad has two shunt and one series. Many of the considerations in the previous section (5.3.2) apply and by similar reasoning it can be shown that:

$$R_1 = R_o\left(\frac{N+1}{N-1}\right) \qquad \text{where} \quad N = \text{antilog}\,\frac{\alpha}{20}$$

$$R_2 = R_o\left(\frac{N^2-1}{2N}\right)$$

and α is the pad attenuation in decibels.

Now let $\dfrac{N+1}{N-1} = n_1$ and $\dfrac{N^2-1}{2N} = n_2$

then Table 5.3 can be used as before to assist in design.

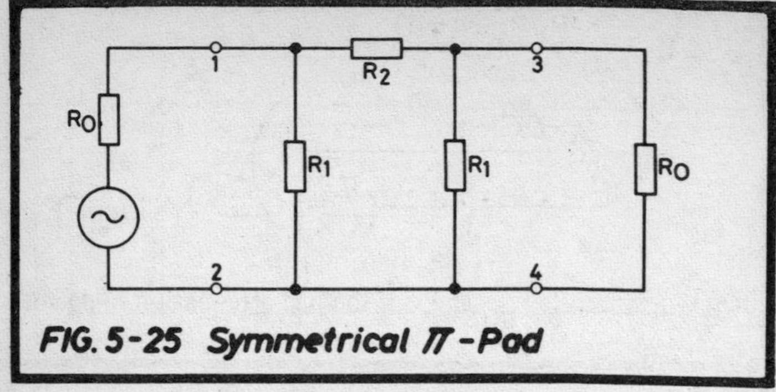

FIG. 5-25 Symmetrical Π-Pad

EXAMPLE:
Re-design the pads shown in Fig.5.22 as symmetrical π-types.

Calculation of pad resistance values: from Table 5.3:

3 dB pad	$n_1 = 5.854$	$\therefore R_1 = 600 \times 5.854 = 3512$ ohms
	$n_2 = .3519$	$\therefore R_2 = 600 \times .3519 = 211$ "
6 dB pad	$n_1 = 3.010$	$\therefore R_1 = 600 \times 3.010 = 1806$ "
	$n_2 = .7469$	$\therefore R_2 = 600 \times .7469 = 448$ "
10 dB pad	$n_1 = 1.925$	$\therefore R_1 = 600 \times 1.925 = 1155$ "
	$n_2 = 1.423$	$\therefore R_2 = 600 \times 1.423 = 854$ "

Using the above values for the two resistors R_1 and for R_2 for each pad, the circuit of Fig.5.22 can be redesigned.

Asymmetrical

The design formulae are more complicated than for the T-network and do not permit the type of simplification that is necessary for a table, such as Table 5.5, to be constructed as an aid to design. Referring to Fig.5.26, the formulae are:

$$R_1 = R_H \left(\frac{N^2 - 1}{N^2 + 1 - 2N \sqrt{\frac{R_H}{R_L}}} \right)$$

$$R_2 = \frac{\sqrt{R_H R_L}(N^2-1)}{2N}$$

$$R_3 = R_L \left(\frac{N^2-1}{N^2+1-2N\sqrt{\frac{R_L}{R_H}}} \right)$$

where $N = \text{antilog} \frac{\alpha}{20}$ and α is the pad attenuation in decibels.

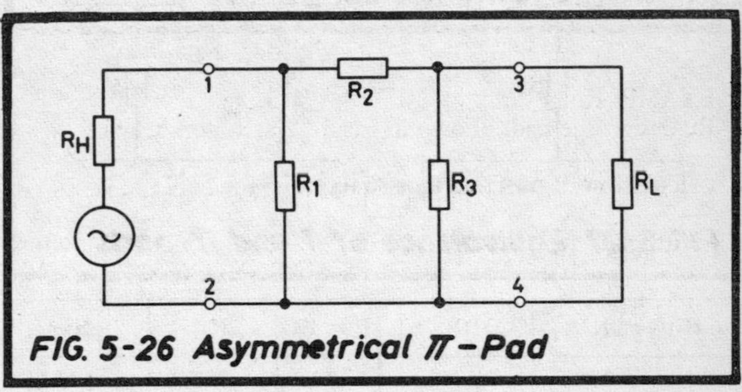

FIG. 5-26 Asymmetrical π-Pad

The values of the pad resistors may therefore be calculated directly from the above, but see the next section for a suggestion of an alternative method of calculation. For direct calculation, values of N and N^2 may be obtained from Table 5.6.

5.3.4 Choice Between T and π-type Networks

The two preceding sections have shown the complete design procedure for both T and π-type networks but no reference has been made as to which is preferable in a particular case. The performance of both is the same, the only differences being configuration and resistance values. Configuration may be important if, for example, d.c. is being carried by the attenuator in addition to its normal function and its resistance to d.c. matters. With resistance value, it may well be that one type will have more suitable values by being nearer to preferred values, than the other.

Therefore it is suggested that design should commence with the T-type as this is simpler, if there remains any doubt about its suitability, then a π-type can be calculated directly from the T design by use of the star-delta transformation, developed in Sect.5.1.6 which also shows the equivalence to T to π transformation.

Thus, having calculated the values R_1, R_2 and R_3 for a T-pad, the equivalent π-pad follows as shown in Fig.5.27.

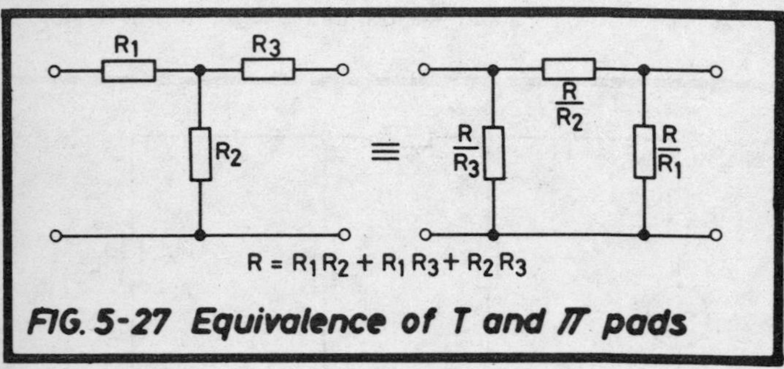

FIG. 5-27 Equivalence of T and π pads

∝, dB	N	N^2	∝, dB	N	N^2
1	1.122	1.259	17	7.079	50.11
2	1.259	1.585	18	7.943	63.09
3	1.413	1.997	19	8.913	79.44
4	1.585	2.512	20	10.00	100.0
5	1.778	3.162	22	12.59	158.5
6	1.995	3.980	24	15.85	251.2
7	2.239	5.013	26	19.95	398.1
8	2.512	6.310	28	25.12	631.0
9	2.818	7.941	30	31.62	1000
10	3.162	10.00	35	56.23	3162
11	3.548	12.59	40	100.0	10,000
12	3.981	15.85	45	177.8	31,623
13	4.467	19.95	50	316.2	10^5
14	5.012	25.12	55	562.3	316,230
15	5.623	31.62	60	1000	10^6
16	6.310	39.82			

N = antilog $\frac{\alpha}{20}$ where α = pad attenuation in decibels.

Table 5.6 **Values of N and N^2 for Asymmetrical π-type Pads**

EXAMPLE:
Re-design the 10 dB T-pad of Fig.5.24 as a π-pad. Use the calculated (not preferred) values.

The T-pad resistance values are

R_1 = 53.16 ohms, R_2 = 38.49 ohms, R_3 = 10.39 ohms

$$R = R_1R_2 + R_1R_3 + R_2R_3 = 2046.13 + 552.33 + 399.91$$

$$= 2998.37$$

$$\frac{R}{R_1} = \frac{2998.37}{53.16} = 56.4 \text{ ohms}$$

$$\frac{R}{R_2} = \frac{2998.37}{38.49} = 77.9 \text{ ohms}$$

$$\frac{R}{R_3} = \frac{2998.37}{10.39} = 288.6 \text{ ohms}$$

The original T-pad and its π equivalent are shown in Fig.5.28.

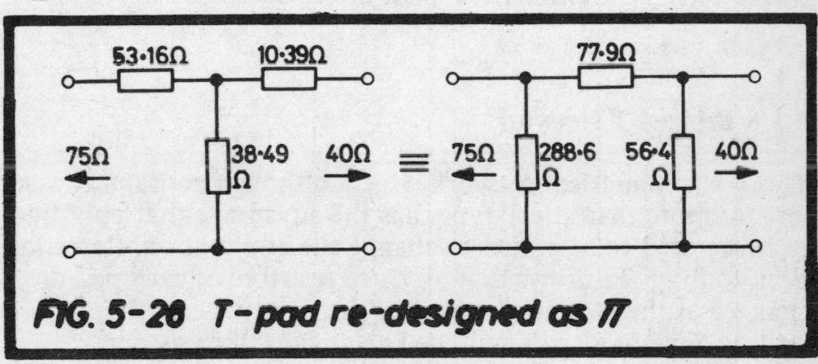

FIG. 5-28 T-pad re-designed as π

5.3.5 Balanced L, T and π-type Networks

The pad series resistance must be equally divided between the two legs of the circuit as shown in Fig.5.29 for the three types.

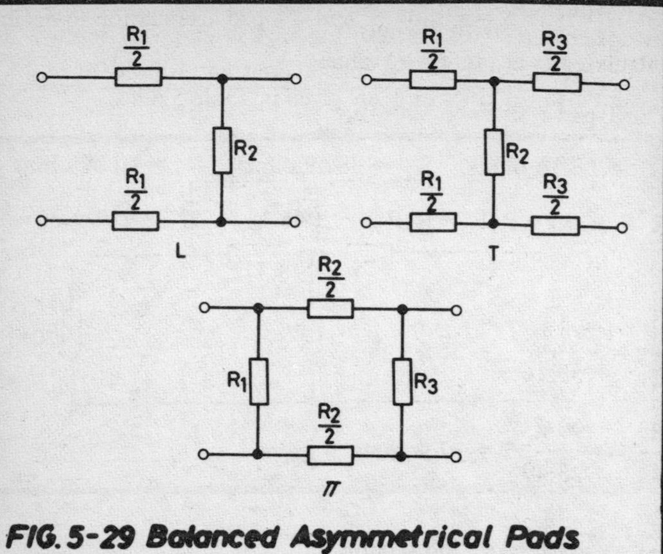

FIG. 5-29 Balanced Asymmetrical Pads

Design proceeds as before, with one further step, i.e. of dividing the series resistance value(s) by 2 and connecting this in each leg. Symmetrical pads are treated similarly.

5.3.6 Bridged-T Network

This is a symmetrical network which, although containing one resistor more than the T-type, has the advantage that only two resistors need to be varied to change the attenuation. Referring back to Fig.5.22 shows that all three resistors of each pad are changed as the switch moves, thus the advantage of the bridged-T network in a switched attenuator increases in proportion to the number of attenuator steps. Fig.5.30 shows the circuit.

Design formulae:

$$R_1 = R_o \quad R_2 = \frac{R_o}{(N-1)} \quad R_3 = R_o(N-1)$$

where $N = \text{antilog} \frac{\alpha}{20}$ and α = pad attenuation in decibels.

Thus the two resistors R_1 remain constant, irrespective of attenuation. Values of N are given in Table 5.6.

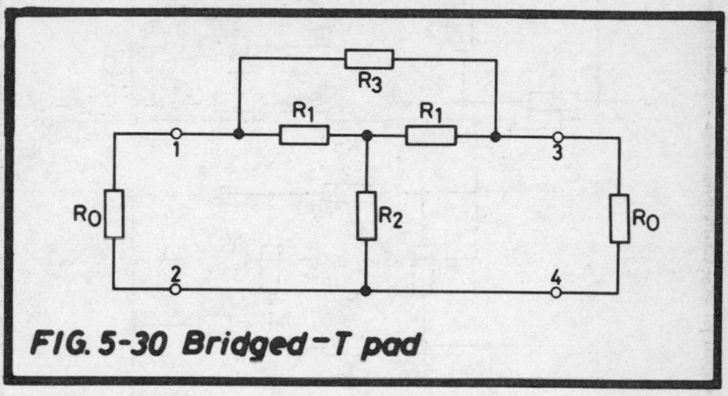

FIG. 5-30 Bridged-T pad

EXAMPLE:
Re-design the circuit of Fig.5.22 using bridged-T networks (3, 6 and 10 dB pads, 600 ohms).

3 dB pad: N (from Table 5.6) = 1.413 (N − 1) = 0.413

R_1 = 600 ohms $R_2 = \frac{600}{0.413}$ = 1452.8 ohms

R_3 = 600 x 0.413 = 247.8 ohms

6 dB pad: N = 1.995 (N − 1) = 0.995

R_1 = 600 ohms $R_2 = \frac{600}{0.995}$ = 603 ohms

R_3 = 600 x 0.995 = 597 ohms

10 dB pad: N = 3.162 (N − 1) = 2.162

R_1 = 600 ohms $R_2 = \frac{600}{2.162}$ = 277.5 ohms

R_3 = 600 x 2.162 = 1297 ohms

The variable 600 ohm attenuator using bridged-T networks is therefore as shown in Fig.5.31.

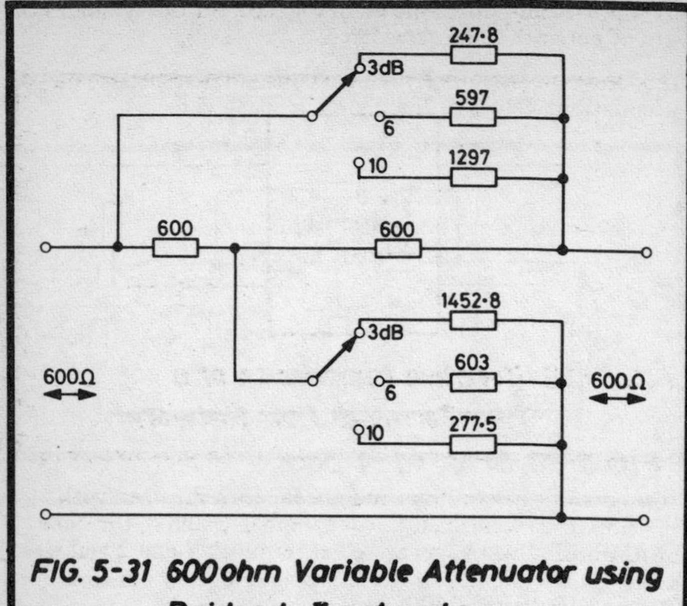

FIG. 5-31 600ohm Variable Attenuator using Bridged–T networks

5.4 FILTERS

The design of very sharp cut-off filters is a subject on its own, warranting a whole text-book for full treatment. It is also unlikely that there will be a call for the more complicated filters, if at all, in the amateur workshop. Nevertheless there are occasions when a simple low or high pass filter is needed, for example, for attenuation of noise in audio systems, therefore design details of these are included here. A few definitions are first appropriate.

Iterative Impedance:
This is defined as "the value of the impedance measured at one pair of terminals of a two-terminal-pair network when the other pair of terminals is terminated with an impedance of the same value."

Diagrammatically this is shown in Fig.5.32 for the symmetrical types of network to be considered.

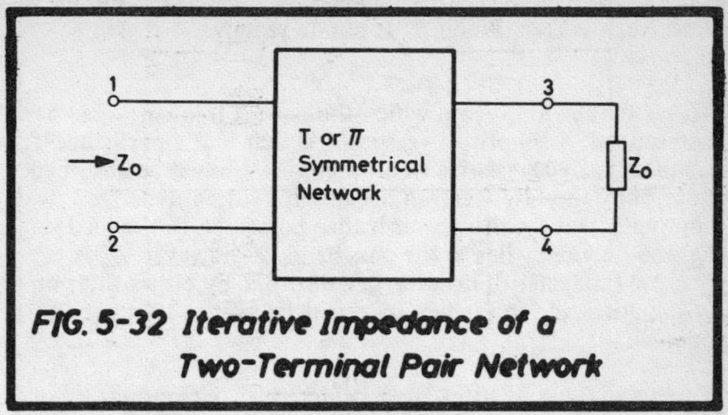

FIG. 5-32 Iterative Impedance of a Two-Terminal Pair Network

If the network has an iterative impedance of Z_o, then with this impedance connected across terminals 3 and 4, the same impedance, Z_o, will be measured at terminals 1 and 2 and vice versa.

As also found in the case of attenuators, the performance of the network depends upon the terminating impedances, and the network must bear the correct relationship to these, they are called the "design impedances" for the particular filter section. When this is done, sections may be connected in series, each being correctly terminated as can be seen from Fig.5.32 where, if the impedance Z_o is replaced by terminals 1 and 2 of another similar network, itself terminated at terminals 3 and 4 by Z_o, the original impedance measured at terminals 1 and 2 of the network shown is unchanged and both the original and the added network are correctly terminated.

Characteristic Impedance:
This is a term more usually associated with transmission lines but may be used with symmetrical attenuator and filter sections. In Fig.5.32 it denotes the common value assumed by the iterative and terminating impedances, i.e. Z_o. In the cases being considered, characteristic impedance is synonymous with iterative impedance.

Variation of Attenuation ($\propto$) with Iterative Impedance Z_o: For a correctly terminated ideal (i.e. no resistance losses) filter section, for the range of frequencies over which Z_o is purely resistive, $\propto$ is zero. When Z_o is purely reactive, $\propto$ is greater than zero.

This is the theorem from which the cut-off frequency can be determined. A simplified reasoning is that if at a particular frequency, Z_o is resistive (real) then power must be absorbed, but not by the filter section because it contains reactive components only, the power is thus passed on to the load, i.e. the frequency lies in the pass band. If, however, Z_o is reactive (imaginary), no power is absorbed by either filter or termination, the frequency therefore lies in the attenuation band.

Two basic types of filter are considered, the "Constant-k" or "Prototype" and the "M-Derived". In the constant-k section the component impedances are related directly to the design impedance, in the m-derived, an improved cut-off characteristic is obtained by changing the internal filter component relationship.

Generally the T-section is considered, but where appropriate the π-section is also shown.

5.4.1 Constant-k Low-Pass Filters

Fig.5.33 shows the elements of the basic sections.

For the constant-k section, the total series and shunt impedances (i.e. of L and C in Fig.5.33) are related to the "design impedance" by

$$Z_1 Z_2 = R_o^2$$

where Z_1 and Z_2 are the total series and shunt impedances respectively and R_o is the design impedance which is made purely resistive.

Hence, from Fig.5.33 $Z_1 = j\omega L \quad Z_2 = \frac{-j}{\omega C}$

and if $Z_1Z_2 = R_o^2$,

$$j\omega L \times \frac{-j}{\omega C} = R_o^2 = \frac{L}{C} \qquad (j^2 = -1)$$

$$\therefore R_o = \sqrt{\frac{L}{C}}$$

– note that this applies to both types of section. Z_o in fact varies with frequency so this design value of R_o is chosen to provide a reasonable compromise.

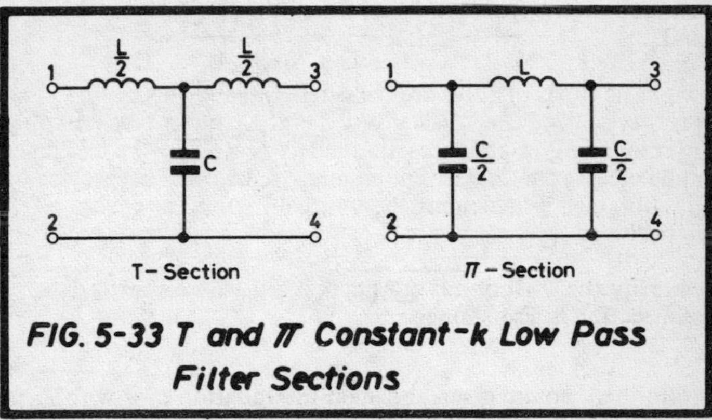

FIG. 5-33 T and Π Constant-k Low Pass Filter Sections

Looking into terminals 1 and 2 of the general T-section of Fig.5.34

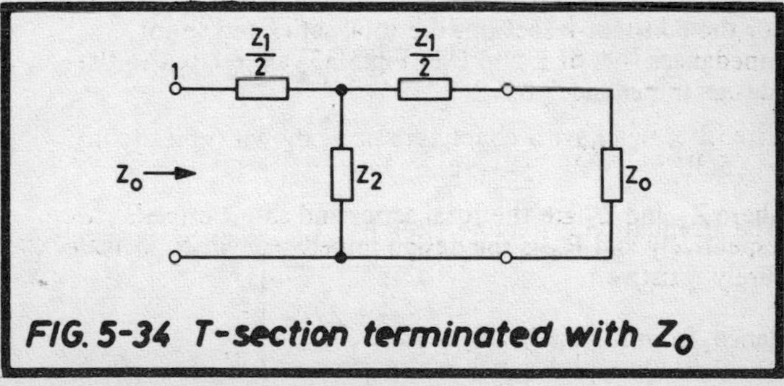

FIG. 5-34 T-section terminated with Z_o

$$Z_o = \frac{Z_1}{2} + \frac{Z_2\left(\frac{Z_1}{2} + Z_o\right)}{Z_2 + \frac{Z_1}{2} + Z_o}$$

from which, $Z_o = \sqrt{\frac{Z_1^2}{4} + Z_1 Z_2}$.

This is the general formula. In terms of the reactive components of Fig.5.33, it becomes:

$$Z_o = \sqrt{\frac{(j\omega L)^2}{4} + j\omega L \times \frac{-j}{\omega C}} = \sqrt{\frac{-\omega^2 L^2}{4} + \frac{L}{C}}$$

$$= \sqrt{1 - \frac{\omega^2 LC}{4}} \times \sqrt{\frac{L}{C}}$$

but $R_o = \sqrt{\frac{L}{C}}$

$$\therefore Z_o = R_o \sqrt{1 - \frac{\omega^2 LC}{4}}$$

From the theorem given above on the variation of $\propto$ with Z_o; the pass band occurs over the range for which Z_o is resistive. This is only when $\omega^2 LC/4 < 1$ and the cut-off frequency, f_c, arises at the change-over point, i.e. when

$$\frac{\omega^2 LC}{4} = 1 \qquad \text{i.e.} \quad f_c = \frac{1}{\pi\sqrt{LC}}$$

The filter will have a characteristic as shown typically in Fig.5.35.

The attenuation $\propto$ may be given by

$$20 \log_{10} \left| 1 + \frac{Z_1}{2Z_2} + \frac{\sqrt{\frac{Z_1^2}{4} + Z_1 Z_2}}{Z_2} \right|$$

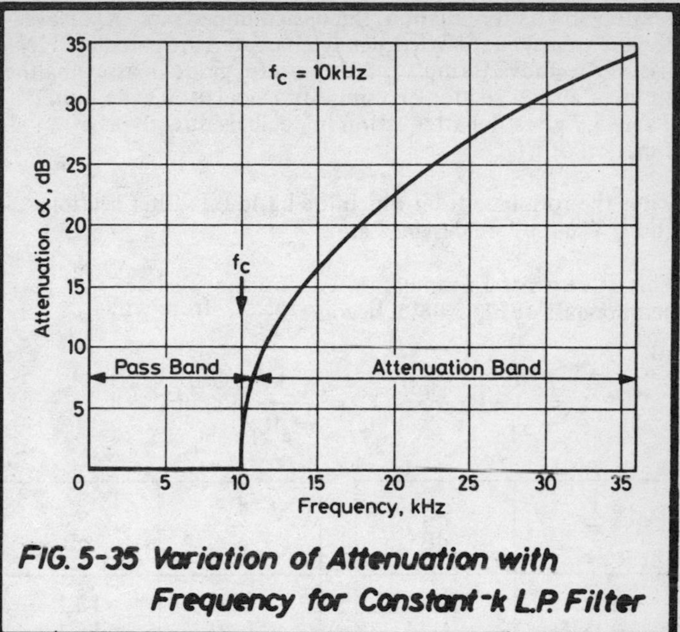

FIG. 5-35 Variation of Attenuation with Frequency for Constant-k L.P. Filter

This is a complicated expression involving much manipulation of the operator j, but a more easily managed expression may be developed by use of hyperbolic functions (the mathematical treatment involves line propagation constants which are not considered here) and this is

$$\propto \; = \; 17.37 \cosh^{-1} \frac{f}{f_c} \text{ dB}$$

where f is the frequency in question and f_c the cut-off frequency.

For two or more sections in series, ∝ is increased proportionately.

It must be emphasized that these formulae are approximate as no account has been taken of resistance in the inductor nor of the frequency dependent mismatch between the filter section and its termination, the design impedance, R_o. Nevertheless an attenuation/frequency characteristic calculated by use of the above formula is a very useful guide in assessing the value of a filter section. To simplify calculation even more, Table 5.7 gives the attenuation in decibels directly as a function of f/f_c.

Thus the formulae from which the basic L.P. filter sections (both T and π) are designed are:

$$R_o = \sqrt{\frac{L}{C}} \quad \text{and} \quad f_c = \frac{1}{\pi\sqrt{LC}} \quad \text{from which,}$$

$$L = \frac{R_o}{\pi f_c} \text{ Henrys,} \quad C = \frac{1}{\pi f_c R_o} \text{ Farads.}$$

$\frac{f}{f_c}$	α, dB	$\frac{f}{f_c}$	α, dB
1.00	0	1.30	13.1
1.005	1.74	1.40	15.1
1.01	2.43	1.50	16.7
1.02	3.47	1.75	20.1
1.03	4.24	2.0	22.9
1.04	4.90	2.5	27.2
1.05	5.47	3.0	30.6
1.06	6.00	4	35.8
1.07	6.46	5	39.8
1.08	6.90	6	43.0
1.09	7.35	7	45.8
1.10	7.71	8	48.1
1.15	9.40	9	50.2
1.20	10.8	10	52.1
1.25	12.0		

f_c = cut-off frequency
f = frequency under consideration.

Table 5.7 **Effect on Attenuation of Deviation from Cut-Off Frequency for Constant-k LP Filter**

EXAMPLE:
A 400 ohm microphone is matched into an amplifier input. A simple LP filter is required to progressively attenuate frequencies above 10 kHz.

$$R_o = 400 \text{ ohms} \qquad f_c = 10{,}000 \text{ Hz}$$

$$\therefore L = \frac{400}{\pi \times 10^4} \text{ H} = 12.73 \text{ mH} \qquad \frac{L}{2} = 6.365 \text{ mH.}$$

$$C = \frac{1}{\pi \times 10^4 \times 400} \text{ F} = 0.0796\,\mu\text{F} \qquad \frac{C}{2} = 0.0398\,\mu\text{F.}$$

The two alternative constant-k sections are shown in Fig.5.36.

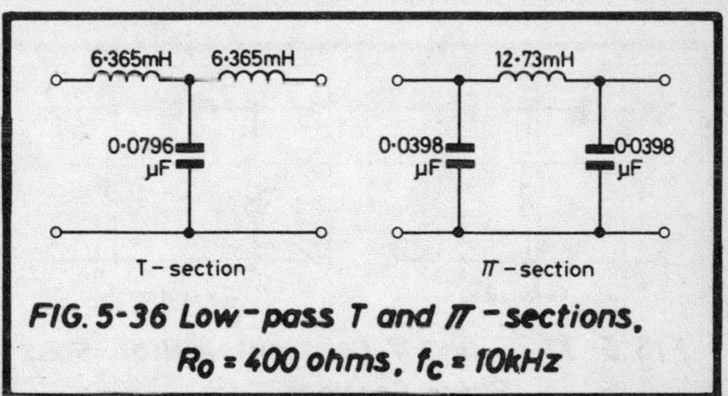

FIG. 5-36 Low-pass T and π-sections, R_o = 400 ohms, f_c = 10kHz

The two sections have similar R_o and f_c values. They differ only in the variation of iterative impedance with frequency.

The attenuation/frequency characteristic is the one actually used for Fig.5.35.

Clearly a filter for this purpose does not need great accuracy in the value of f_c and it is instructive to examine the change if, for example, only 6 mH inductors were available and, as is likely, only preferred values for C could be used.

If L is smaller than the design value, then to maintain R_o, C should also be smaller e.g. .075 μF.

$$\text{Then } f_c = \frac{1}{\pi\sqrt{LC}} = 10{,}610 \text{ Hz} \quad R_o = \sqrt{\frac{L}{C}} = 400 \text{ ohms.}$$

Conversely, should the accuracy of f_c be more important, the next higher preferred value for C would be chosen so that $\sqrt{LC}$ would change least.

Some help with the design of small inductors may be gained from Sect.3.4.2 and with the measurement of inductance from Sect.6.2.3.

5.4.2 Constant-k High Pass Filters

Fig.5.37 shows the elements of the basic sections.

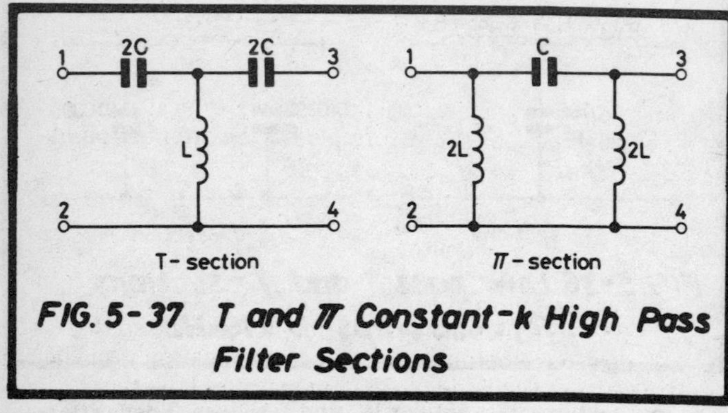

FIG. 5-37 T and π Constant-k High Pass Filter Sections

Being a constant-k section, as in the case of the L.P. filter,

$$Z_1 Z_2 = R_o^2$$

where Z_1 and Z_2 are the total series and shunt impedances and R_o is the (resistive) design impedance.

Hence, from Fig.5.37

$$Z_1 = \frac{-j}{\omega C} \quad Z_2 = j\omega L \quad \text{and} \quad R_o = \sqrt{\frac{L}{C}} \quad \text{for both}$$

T and π-sections.

A similar algebraic approach as for the L.P. filter gives

$$Z_o = R_o \sqrt{1 - \frac{1}{4\omega^2 LC}}$$

where Z_o is the impedance looking into terminals 1 and 2 with terminals 3 and 4 closed with the same value.

When $4\omega^2 LC > 1$, Z_o is real and f is within the pass band, when $4\omega^2 LC < 1$, Z_o is imaginary and f is within the attenuation band.

Hence the cut-off frequency, f_c, is given by

$$4\omega^2 LC = 1 \qquad \therefore\ f_c = \frac{1}{4\pi\sqrt{LC}}$$

The filter will have a characteristic as shown typically in Fig.5.38, calculated from the formula for the attenuation of the H.P. section,

$$\alpha = 17.37 \cosh^{-1} \frac{f_c}{f} \text{ dB.}$$

Fig.5.38 has been plotted for f_c = 10 kHz, the same as used for Fig.5.35. Note that there is a difference in scale so that Fig.5.35 may show better the change in attenuation near the cut-off frequency while Fig.5.38 demonstrates how the attenuation curve rises towards infinity at frequencies very remote from cut-off.

For the H.P. filter, Table 5.8 is included for quick estimation of the attenuation/frequency characteristic.

Rearranging the formulae already given for R_o and f_c, the basic H.P. filter section component values (for both T and π sections) are given by

$$L = \frac{R_o}{4\pi f_c} \text{ Henrys,} \quad C = \frac{1}{4\pi f_c R_o} \text{ Farads.}$$

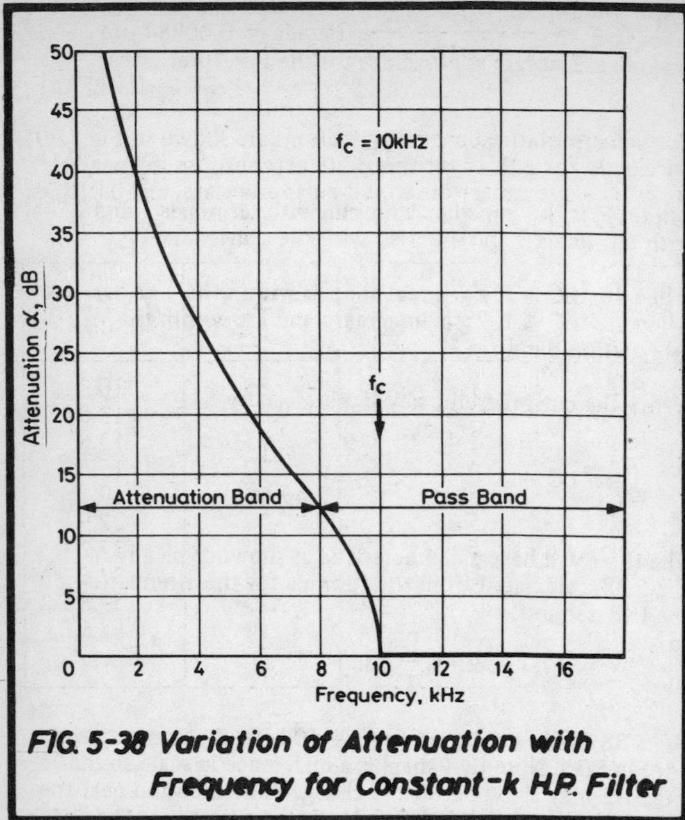

FIG. 5-38 Variation of Attenuation with Frequency for Constant-k H.P. Filter

EXAMPLE:

A 50 kHz supply from a tape recording bias oscillator needs a high pass filter cutting off at about 45 kHz to minimize pick-up of lower frequency signals. Design a constant-k filter for this purpose for connexion between 200 ohm impedances. What is the approximate filter attenuation at 42 kHz?

$$R_o = 200 \text{ ohms}, \quad f_c = 45 \text{ kHz}$$

$$\therefore L = \frac{200}{4\pi \times 45 \times 10^3} \text{ Henrys} = 354\,\mu\text{H} \quad \therefore 2L = 708\,\mu\text{H}$$

$$C = \frac{1}{4\pi \times 45 \times 10^3 \times 200} \text{ Farads} = 0.00884\ \mu F$$

$$\therefore\ 2C = 0.0177\ \mu F$$

The two alternative constant-k sections are shown in Fig.5.39, but clearly for a filter for this particular purpose the nearest preferred value capacitors would be appropriate, e.g. 0.018 and 0.0091 μF.

$\frac{f}{f_c}$	$\propto$, dB	$\frac{f}{f_c}$	$\propto$, dB
.995	1.74	.85	10.2
.990	2.43	.80	12.0
.98	3.47	.75	13.8
.97	4.24	.70	15.6
.96	5.02	.60	19.1
.95	5.63	.50	22.9
.94	6.18	.40	27.2
.93	6.69	.30	32.6
.92	7.19	.20	39.8
.91	7.69	.10	52.1
.90	8.11		

f_c = cut-off frequency
f = frequency under consideration.

Table 5.8 **Effect on Attenuation of Deviation from Cut-Off Frequency for Constant-k H.P. Filter**

Filter attenuation at 42 kHz:–

$$\frac{f}{f_c} = \frac{42}{45} = 0.933 \quad \text{From Table 5.8} \quad \propto = 6.69 \text{ dB for } \frac{f}{f_c} = 0.93$$

i.e. approximate attenuation at 42 kHz = 7 dB.

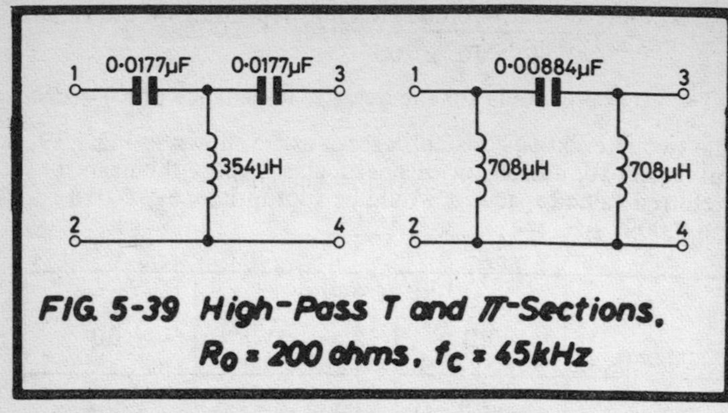

FIG. 5-39 High-Pass T and π-Sections.
R_0 = 200 ohms, f_C = 45kHz

Choice of T or π-section might depend on d.c. considerations, for example, if d.c. has to flow between terminals 1 and 2 or between 3 and 4, the π-section has the obvious advantage. Choice is also affected by availability of components.

5.4.3 M-Derived Filters – General Theory

The "m-derived" filter section has an improved (i.e. steeper) attenuation/frequency characteristic compared with the constant-k section, the price which has to be paid is a more complicated filter in that it contains one more component and actually has less attenuation at frequencies well above the cut-off value, f_c.

The main technical difference between the two types is that the m-derived contains a resonant circuit such that resonance produces (theoretically) infinite attenuation at some desired frequency whereas the constant-k provides infinite attenuation only at zero or infinite frequency according to the type of filter (see Figs.5.35 and 5.38). Both filters have their own particular field of use, this will become evident from the practical examples given later in this section.

Consider first the general form of a rudimentary m-derived section as in Fig.5.40.

In (a) the series impedances $Z_1/2$ of the constant-k section have been replaced by impedances of value $mZ_1/2$, and Z_2

must therefore have some new value depending on the value of m. Let this be $Z_{2(m)}$.

The general formula for the iterative impedance of a T-section which is

$$Z_o = \sqrt{\frac{Z_1^2}{4} + Z_1 Z_2} \quad \text{now becomes}$$

$$Z_o = \sqrt{\frac{m^2 Z_1^2}{4} + m Z_1 Z_{2(m)}}$$

Equating the two expressions shows that

$$Z_{2(m)} = \frac{Z_2}{m} + Z_1\left(\frac{1-m^2}{4m}\right),$$

that is, two separate impedances in series as shown in Fig.5.40(b). Similarly the π-section can be developed as in (c).

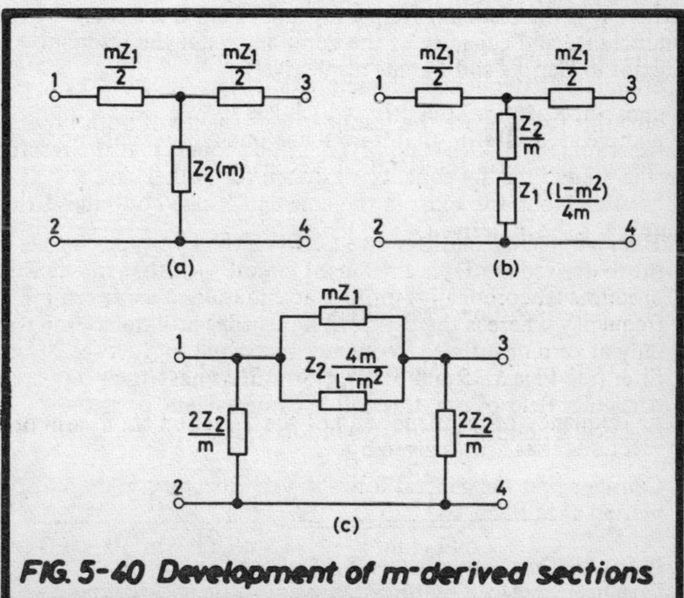

FIG. 5-40 Development of m-derived sections

Referring to the T-section, it will be seen that the new shunt arm contains components of both Z_1 and Z_2, that is, reactances of opposite sign and it is this feature from which the resonance effect is obtained, because there will be some frequency at which the net reactance of the shunt arm is zero, producing series resonance with theoretically zero impedance. At this frequency the attenuation will be infinite because there is a short-circuit across the through path.

Equally in the case of the π-section, a parallel resonant circuit is inserted in series with the through path, at resonance producing infinite impedance and therefore infinite attenuation. Resonance is considered fully in Section 4.7 in which it is shown that, with very little sacrifice of accuracy, both series and parallel resonant circuits containing the same values of inductance and capacity have the same resonant frequency which is given by the general expression

$$f_r = \frac{1}{2\pi\sqrt{LC}}$$

where f_r is the resonant frequency and L and C are the inductance and capacity of the components of the resonant circuit in Henrys and Farads respectively.

Using this to calculate the resonant frequencies of both types of section (Fig.5.40(b) and (c)) it will be found that the result is the same, thus the equality of design for both T and π-sections found to exist in the constant-k case continues to apply for the m-derived.

5.4.4 M-derived Low Pass Filters

The two m-derived sections are shown in Fig.5.41.

The frequency of resonance (f_r) of the T-section shunt arm or π-section series arm is given by

$$f_r = \frac{1}{2\pi\sqrt{\frac{m(1-m^2)}{4m}\cdot LC}} = \frac{1}{\pi\sqrt{(1-m^2)}\times\sqrt{LC}}$$

but since the cut-off frequency, $f_c = \dfrac{1}{\pi\sqrt{LC}}$ (Sect.5.4.1)

$$f_r = \frac{f_c}{\sqrt{1-m^2}} \quad \text{and} \quad \frac{f_r}{f_c} = \frac{1}{\sqrt{1-m^2}}$$

$$\therefore\ m = \sqrt{1-\left(\frac{f_c}{f_r}\right)^2}$$

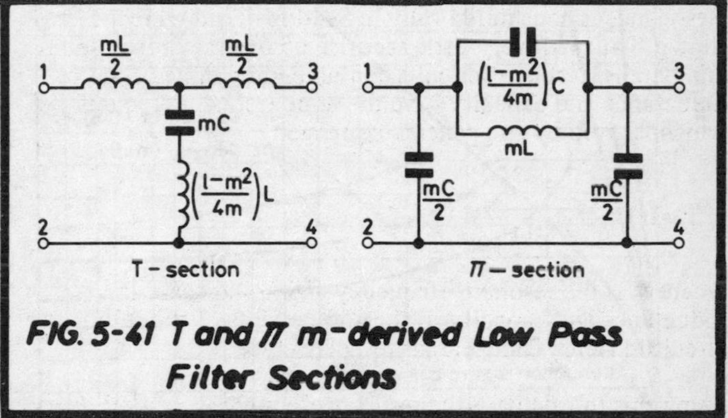

FIG. 5-41 T and Π m-derived Low Pass Filter Sections

The problem, is, of course, what value of m to choose, because although small values of m give sharp cut-off, the attenuation falls at frequencies above this value. This is shown in Figs.5.42(i) and (ii) which show the comparison between the m-derived section and the prototype or constant-k section (for which m = 1). A study of the range of m-derived characteristics as shown in Fig.5.42 with the constant-k characteristic for comparison, and with a knowledge of the purpose for which the filter is required is helpful in making a choice. It must be emphasized that the curves are calculated on the basis of loss-free components and correct matching between the section and its terminations. Although neither condition applies in practice it can be shown that the inaccuracy arising from the assumptions can be made quite small.

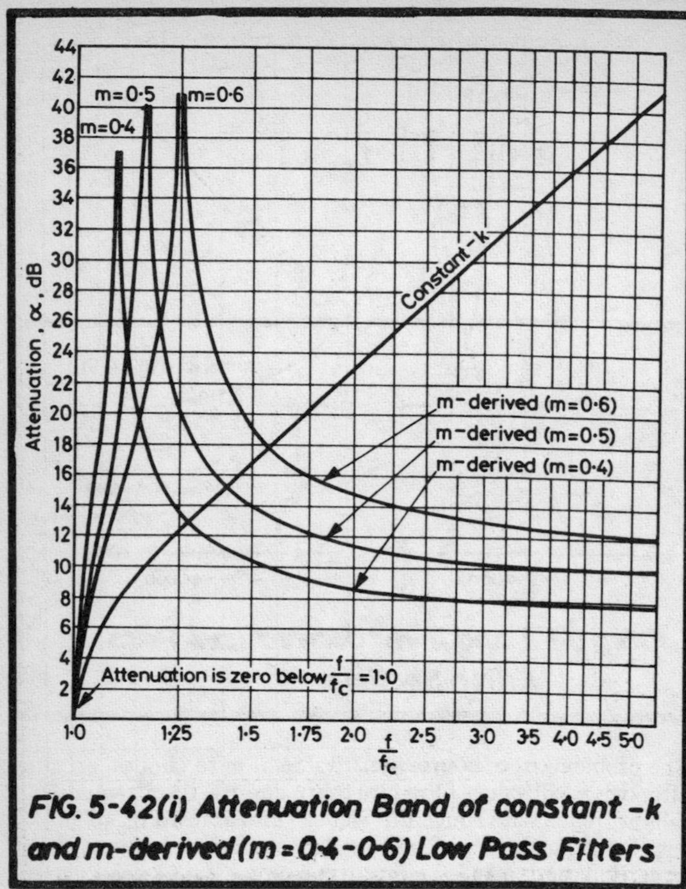

FIG. 5-42(i) Attenuation Band of constant -k and m-derived (m=0·4-0·6) Low Pass Filters

The attenuation/frequency characteristics have been calculated from the general formula for the propagation constant (γ) of a T-section:

$$\cosh \gamma = 1 + \frac{1}{2}\left(\frac{Z_1}{Z_2}\right) \text{ nepers,}$$

and because reactances only are being considered, Z_1 and Z_2 become X_1 and X_2 and the propagation constant will be found

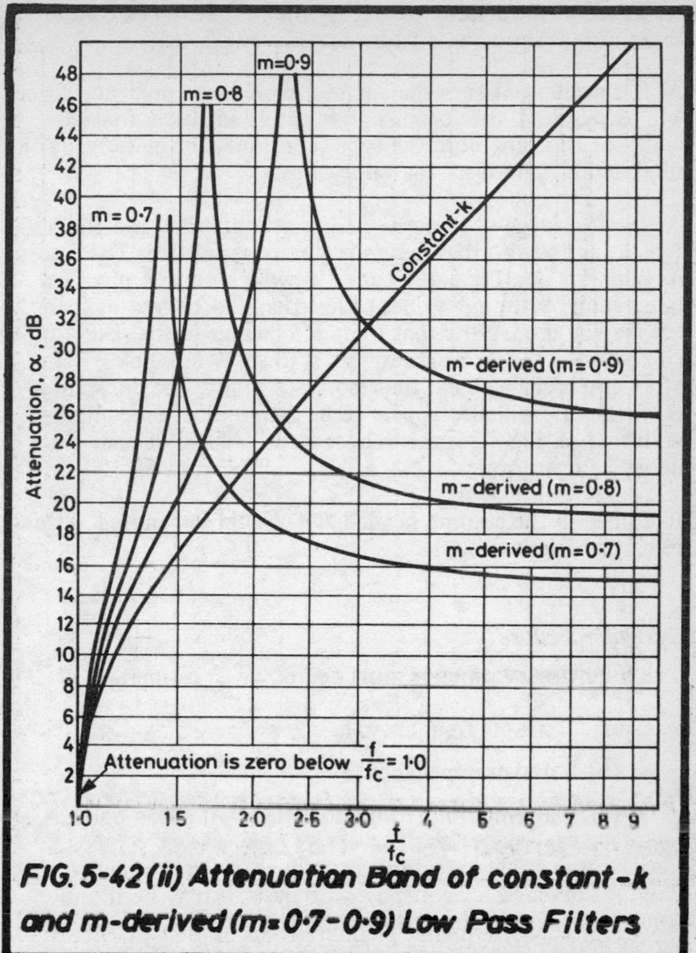

FIG. 5-42 (ii) Attenuation Band of constant-k and m-derived (m=0·7–0·9) Low Pass Filters

to have no phase angle. Hence, for this purpose the formula may be written,

$$\cosh \alpha = 1 + \frac{1}{2}\left(\frac{X_1}{X_2}\right) \text{nepers,}$$

where, for example, in the case of the T-section, X_1 is the reactance of the series arm and X_2 the net reactance of the shunt arm as shown in a later example (Fig.5.45).

As it is unlikely that in the amateur workshop, precision filters will be required, the formula gives ample accuracy for most work for choosing both the type (constant-k or m-derived) of filter and if m-derived, the value of m.

A practical range of m-derived filter attenuations and characteristics is given so that design is greatly simplified. These are in Table 5.9 and Fig.5.42(i) and (ii), which cover a practical range for m, although values of less than 0.4 or greater than 0.9 may be used. Note that there is a change in the scale for f/f_c between the two figures, this is to show more clearly the attenuation change near the resonance point. The constant-k characteristic will also appear to be different when compared with that on Fig.5.35, this is because of change of scale from linear to logarithmic.

In Table 5.9 the column headed f_c/f should be ignored – this refers to high-pass filters.

Design Procedure

The following parameters must be known or estimated:

(i) cut-off frequency, f_c

(ii) design impedance, R_o

(iii) attenuation required in the attenuation band, α.

Because universal characteristics such as shown in Fig.5.42 can only be drawn on a frequency ratio basis, it may be found convenient first to construct a scale converting f/f_c back into actual frequencies relative to the cut-off value required.

The first choice to be made is between the constant-k and m-derived filters. This choice is made straightforward by consideration of Fig.5.42. Assuming an m-derived section is needed the range of characteristics shown will further help in choice of m. If the value chosen happens to be one of those shown, then the actual attenuations at various ratios of f/f_c may be obtained directly from Table 5.9.

$\frac{f}{f_c}$	m						$\frac{f_c}{f}$
	0.4	0.5	0.6	0.7	0.8	0.9	
1.05	17.4	12.3	9.7	8.1	7.0	6.1	0.95
1.10	33.8	20.8	14.9	11.9	10.0	8.7	0.91
1.15	19.6	44.4	20.3	15.3	12.5	10.7	0.87
1.20	15.9	26.0	27.7	18.6	14.7	12.4	0.83
1.25	14.0	20.8	∞	22.3	16.9	14.0	0.80
1.30	12.8	18.3	30.0	26.8	19.0	15.4	0.77
1.35	11.9	16.7	25.8	33.8	21.2	16.8	0.74
1.40	11.3	15.6	22.3	80.0	23.5	18.1	0.71
1.45	10.8	14.8	20.6	35.4	26.1	19.3	0.69
1.50	10.4	14.1	19.3	30.1	29.0	20.5	0.67
1.6	9.8	13.2	17.7	25.3	38.2	23.0	0.63
1.7	9.4	12.5	16.6	22.9	45.4	25.4	0.59
1.8	9.1	12.1	15.8	21.3	34.3	28.1	0.56
1.9	8.9	11.7	15.3	20.3	30.3	30.9	0.53
2.0	8.7	11.4	14.8	19.5	28.0	34.3	0.50
2.2	8.4	11.0	14.2	18.4	25.4	45.7	0.45
2.4	8.2	10.7	13.8	17.7	23.9	46.0	0.42
2.6	8.1	10.5	13.5	17.2	22.9	38.0	0.38
2.8	8.0	10.4	13.2	16.9	22.2	34.6	0.36
3.0	7.9	10.3	13.1	16.6	21.7	32.7	0.33
3.5	7.7	10.1	12.8	16.2	20.9	30.1	0.29
4.0	7.6	9.9	12.6	15.9	20.4	28.8	0.25
5.0	7.5	9.8	12.4	15.6	19.9	27.4	0.20
7.0	7.4	9.7	12.2	15.3	19.5	26.5	0.14
10.0	7.4	9.6	12.1	15.2	19.3	26.0	0.10
$\frac{f}{f_c}$ at f_r	1.091	1.155	1.250	1.400	1.667	2.294	← LOW PASS
HIGH PASS →	0.917	0.866	0.800	0.714	0.600	0.436	$\frac{f_c}{f}$ at f_r

f_c = frequency of cut-off
f_r = frequency of resonance.

Table 5.9 **M-derived Filter Attenuations in Decibels**

Design of the filter follows from the formulae for the basic (constant-k) section i.e.

$$L = \frac{R_o}{\pi f_c} \text{ Henrys} \qquad C = \frac{1}{\pi f_c R_o} \text{ Farads.}$$

and, for example, for the T-section, as shown in Fig.5.41, the series inductances are of value mL/2 and the shunt arm has a capacitor of value mC in series with an inductor $(1-m^2/4m)L$. The π-section component values are calculated similarly from Fig.5.41.

This, in fact, completes the paper design of the filter and its attenuation/frequency characteristic is as shown on Fig.5.42 for the appropriate value of m.

If values of m shown are not suitable, for example there is a precise frequency at which very high attenuation is required, then the value of m must first be calculated. The frequency of resonance, f_r is the one at which attenuation is highest, and the value of m is linked to this as already shown by

$$m = \sqrt{1 - \left(\frac{f_c}{f_r}\right)^2}$$

By artistic interpolation on Fig.5.42, the characteristic may be sketched, if however, greater accuracy is required, recourse may be had to the formula already given for calculating the attenuation either directly with frequency or the ratio f/f_c.

The following examples should amplify and clarify the procedure.

EXAMPLE:
A single section low-pass filter is required with as steep a cut-off at 10 kHz as possible and with at least 15 dB attenuation at 100 kHz. It is to work between 200 ohm impedances. Design the filter.

This is quite straightforward because Fig.5.42(ii) shows that the m-derived section with m = 0.7 has an attenuation of 15.2 dB (Table 5.9) at $f/f_c = 10$, i.e. 100 kHz.

(i) $f_c = 10$ kHz $\qquad m = 0.7 \qquad \dfrac{1-m^2}{4m} = 0.182$

(ii) $R_o = 200$ ohms

T-section:

$$L = \frac{R_o}{\pi f_c} H = \frac{200}{\pi \times 10^4} = 6.366 \text{ mH}$$

$$\therefore \qquad \frac{mL}{2} = \frac{0.7 \times 6.366}{2} = 2.228 \text{ mH}$$

$$\left(\frac{1-m^2}{4m}\right)L = 0.182 \times 6.366 = 1.159 \text{ mH}$$

$$C = \frac{10^6}{\pi f_c R_0} \mu F = \frac{10^6}{\pi \times 10^4 \times 200} = 0.1592 \mu F$$

$$\therefore \qquad mC = 0.7 \times 0.159 = 0.1114 \mu F$$

π-section:

From T-section, $L = 6.366$ mH

$$\therefore \qquad mL = 0.7 \times 6.366 = 4.456 \text{ mH}$$

$C = 0.1592 \mu F$

$$\therefore \qquad \frac{mC}{2} = \frac{0.7 \times 0.1592}{2} = 0.0557 \mu F$$

$$\left(\frac{1-m^2}{4m}\right)C = 0.182 \times 0.1592 = 0.029 \mu F$$

The filter sections are illustrated in Fig.5.43. Either can be used.

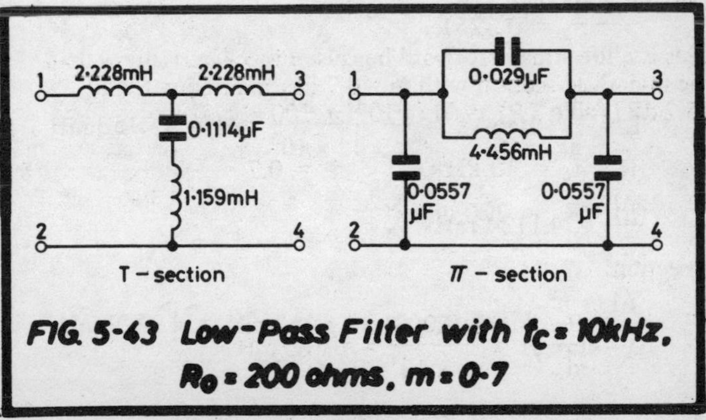

FIG. 5-43 Low-Pass Filter with f_c = 10kHz, R_0 = 200 ohms, m = 0·7

EXAMPLE:
In a system of ultrasonic alarms, one unit working at 40 kHz is occasionally triggered falsely by other remote units, especially one working at 50 kHz. Design a low pass filter for the 40 kHz unit to install between the transducer and the amplifier input (both 2000 ohms) to have as high an attenuation as possible at 50 kHz.

The requirement may be satisfied by an m-derived low-pass filter having a cut-off frequency slightly above 40 kHz, say at 42 kHz with such a value of m that highest attenuation is obtained at 50 kHz.

Thus: $f_c = 42$ kHz

$f_r = 50$ kHz $\qquad \dfrac{f_r}{f_c} = 1.1905$

$R_o = 2000$ ohms

The bottom line of Table 5.9 shows that $f_r/f_c = 1.1905$ falls between m = 0.5 and 0.6, thus a vertical line could be drawn on Fig.5.42(i) at $f/f_c = 1.19$ (the line which the characteristic theoretically meets at infinity), the characteristic sketched in and the estimated attenuations read off.

The filter design, say for a T-section, is as follows:

$$m = \sqrt{1 - \left(\frac{f_c}{f_r}\right)^2} = \sqrt{1 - \left(\frac{42}{50}\right)^2} = 0.5426$$

$$\frac{1 - m^2}{4m} = 0.3251$$

$$L = \frac{R_o}{\pi f_c} \text{ Henrys} = \frac{2000 \times 10^3}{\pi \times 42 \times 10^3} \text{ mH} = 15.158 \text{ mH}$$

$$\frac{mL}{2} = 4.1124 \text{ mH}$$

$$\left(\frac{1 - m^2}{4m}\right) L = 0.3251 \times 15.158 \text{ mH} = 4.9279 \text{ mH}$$

$$C = \frac{10^6}{\pi f_c R_0} = \frac{10^6}{\pi \times 42 \times 10^3 \times 2000} \mu F = .003789 \mu F$$

$$mC = 0.5426 \times .003789 = .002056 \mu F$$

The filter section is shown in Fig.5.44.

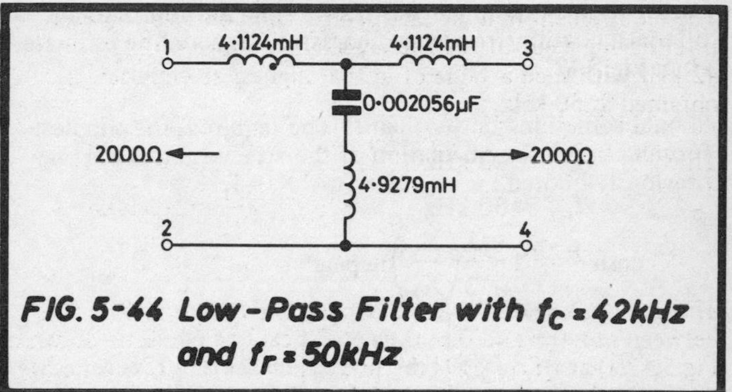

FIG. 5-44 Low-Pass Filter with f_c = 42kHz and f_r = 50kHz

This is the correct filter and it will be found that the shunt arm resonates at almost exactly 50 kHz, irrespective of resistance losses. But the values of the components may be difficult to realise in practice.

Inductors may be overwound when constructed and turns removed until the correct value of inductance is measured but capacitors are usually bought in preferred values, not made. Thus minor changes to component values may be made to obtain more practical values, e.g. if in the shunt arm the capacitance is made .002 µF then with the inductance raised slightly to 5.066 mH, the arm will still resonate at 50 kHz, which is the most important function of the filter. This calculation follows from the formula for the frequency of resonance of the shunt arm

$$f_r = \frac{1}{2\pi\sqrt{LC}}$$ where L and C are the components of the shunt arm in Henrys and Farads respectively

$$\therefore L = \frac{1}{(2\pi f_r)^2 C} = \frac{10^6}{4\pi^2 \times 25 \times 10^8 \times .002} \text{ H} = 5.066 \text{ mH}$$

(However, it just so happens that the exact value of capacitance in this particular case would be relatively easy to provide by use of a .002 μF in parallel with a 56 pF).

Earlier in this example it was shown how the approximate filter attenuation/frequency characteristic could be estimated from Fig.5.42.

Should something better than this be required, the simplest formula to use for calculation of the attenuation is the one previously quoted i.e.

$$\cosh \alpha = 1 + \frac{1}{2}\left(\frac{X_1}{X_2}\right) \text{nepers}$$

$$= 8.686\left[1 + \frac{1}{2}\left(\frac{X_1}{X_2}\right)\right] \text{decibels} \qquad \text{(Sect.6.2.6)}$$

where α is the filter attenuation and X_1 and X_2 are the net reactances of the filter sections as shown in Fig.5.45.

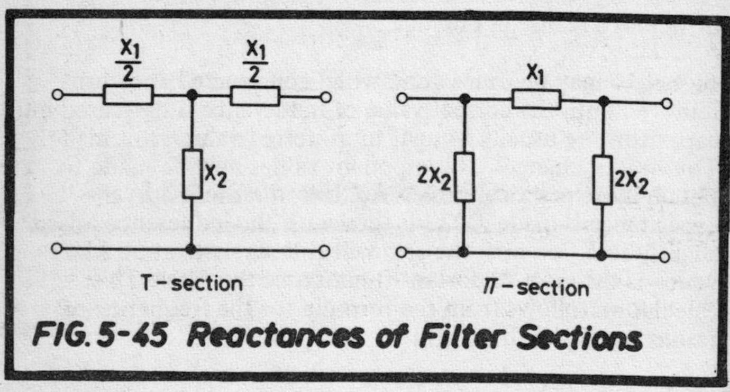

FIG. 5-45 Reactances of Filter Sections

EXAMPLE:
Transform the low-pass T-section filter shown in Fig.5.36 into an m-derived section with m = 0.6. Show by calculation that

the filter has minimum attenuation at 10 kHz, maximum at 12.5 kHz. Also calculate the attenuation at 10.75 and 21 kHz.

For the m-derived sections:

Series arm inductance = mL = 0.6 x 12.73 = 7.638 mH

$$\text{Shunt arm inductance} = \left(\frac{1-m^2}{4m}\right)L = 0.2667 \times 12.73$$

$$= 3.395 \text{ mH}$$

Shunt arm capacitance = mC = 0.6 x 0.0796 = 0.04776 μF

as shown for the T-section in Fig.5.46.

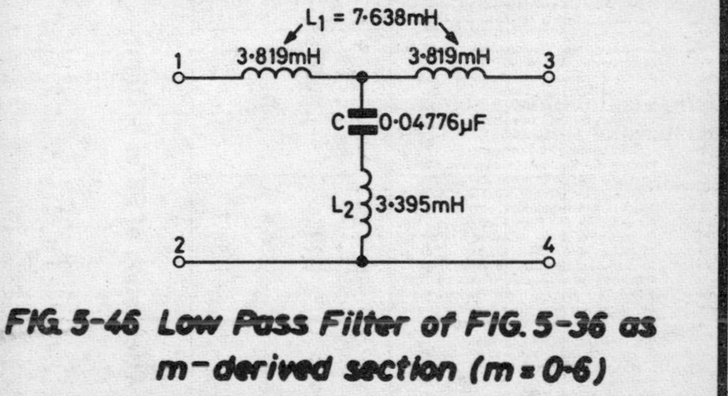

FIG. 5-46 Low Pass Filter of FIG. 5-36 as m-derived section (m = 0·6)

1	2	3	4	5	6	7	8	9
f_1 kHz	X_1 $= 47.99 \times$ Col.1	X_{L_2} $= j21.33 \times$ Col.1	X_C $= \frac{-j3332.39}{\text{Col.1}}$	X_2 = Col.3 + Col.4	$\frac{1}{2}\left(\frac{X_1}{X_2}\right)$ $= \frac{1}{2}\frac{\text{Col.2}}{\text{Col.5}}$	$1 + \frac{1}{2}\left(\frac{X_1}{X_2}\right)$ = 1 + Col.6	$\cosh^{-1} 1 + \frac{1}{2}\frac{X_1}{X_2}$ $= \cosh^{-1}$ Col.7 nepers	α $= 8.686 \times$ Col.8 dB
10	j 479.9	j213.3	−j333.24	−j119.9	−2.0	−1.0	0	0
10.75	j 515.89	j229.3	−j310.0	−j 80.7	−3.196	−2.196	1.423	12.4
12.5	j 599.88	j266.6	−j266.6	0	∞	∞	∞	∞
21	j1007.8	j447.9	−j158.7	j289.2	1.742	2.742	1.667	14.5

Table 5.10 **Calculation of Attenuation of Filter Section**

When several calculations from the same formula are required it is often convenient to first design a table so that columns may be calculated separately, thus minimizing the risk of error. An example follows:

X_1 (Figs.5.45 and 5.46) $= 2\pi fL_1 = 47.99 \times f$ where f is in kHz

X_{L_2} " " $= 2\pi fL_2 = 21.33 \times f$ " " " "

X_C " " $= \dfrac{-1}{\omega C} = \dfrac{-3332.39}{f}$ " " " "

These figures simplify the repeated calculations as typically shown for this example in Table 5.10. The figures in Column 9 give the answers required.

Notes on Table 5.10:

1. Values of cosh x are given in most books of tables.
2. The attenuation values calculated can be checked from Fig.5.42.
3. It will be observed that the operator j can be omitted because it ultimately cancels out in column 6.
4. The signs in column 7 are irrelevant to the calculation of α, the negative sign simply indicating a phase change of 180° through the filter.

5.4.5 M-derived High-Pass Filters

There is such similarity between the low and high-pass m-derived filters in their analysis and design that this section should be treated more as a supplement to the previous section than as a complete design procedure on its own.

The T and π m-derived high-pass filter sections are shown in Fig.5.47.

The overall series and shunt component values L and C are calculated from the design formulae for the constant-k or prototype section i.e.

$$L = \frac{R_0}{4\pi f_c} \text{ Henrys} \qquad C = \frac{1}{4\pi f_c R_0} \text{ Farads}$$

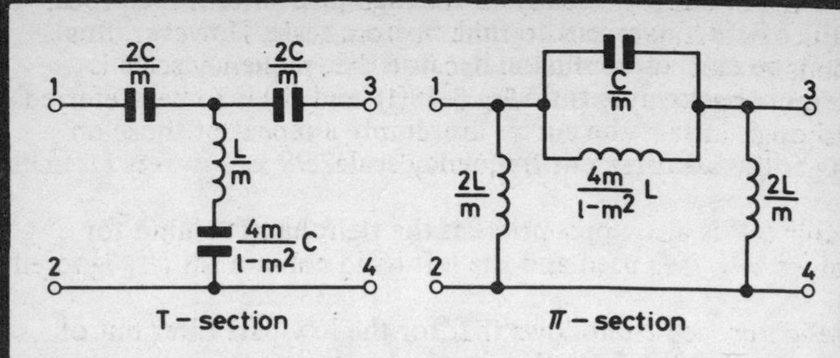

FIG. 5-47 T and π m-derived High-Pass Filter Sections

With L and C determined, the component values in the circuits of Fig.5.47 follow when the value of m is known.

As in the case of the low-pass filter, infinite attenuation is theoretically obtained in the T-section when resonance occurs in the shunt arm and in the π-section when it occurs in the series arm and these are at the same frequency, f_r. For either section:

$$f_r = \frac{1}{2\pi\sqrt{\frac{4}{1-m^2}LC}} = \frac{1}{4\pi\sqrt{\frac{LC}{1-m^2}}}$$

but since the cut-off frequency, $f_c = \dfrac{1}{4\pi\sqrt{LC}}$ (Sect.5.4.2)

$$f_r = f_c\sqrt{1-m^2} \quad \text{and} \quad \frac{f_r}{f_c} = \sqrt{1-m^2}$$

$$\therefore\ m = \sqrt{1-\left(\frac{f_r}{f_c}\right)^2}$$

It will be noticed that, compared with the low-pass case, the relationship of m with f/f_c has become reciprocal, thus the attenuation/frequency characteristics for various values of m of Fig.5.42 apply equally to the high-pass filter if reciprocal values for f/f_c are used on the bottom scale. However, this tends to lead to confusion because the frequency scale is running backwards thus Fig.5.48(i) and (ii) is to be preferred although in fact the curves are simply a repeat of those on Fig.5.42 to a different frequency scale.

Table 5.9 is also appropriate if the right-hand column for values of f_c/f is used and the left-hand column for f/f_c ignored.

Design procedure follows that for the low-pass filter but of course using the formulae for the high-pass.

EXAMPLE:
It is required to filter out two audio tones at 1200 and 1600 Hz on a 600 ohm line carrying higher frequency signals. At least 30 dB attenuation is required at 1200 Hz and 12 dB at 1600 Hz, with minimum attenuation at and above 2400 Hz. Design a suitable single-section filter.

A high-pass filter is required and first, f_c has to be determined, it must not be higher than 2,400 Hz, preferably slightly lower.

Consider the values of f/f_c for 1200 and 1600 Hz for various values of f_c

f_c, Hz	$\frac{f}{f_c}$ at 1200 Hz	$\frac{f}{f_c}$ at 1600 Hz
2400	0.5	0.67
2200	0.55	0.73
2000	0.6	0.8

With these ratios in mind, Fig.5.48 can be studied and it is firstly evident that a constant-k filter cannot provide the attnuation required because this does not reach 30 dB unless the ratio f/f_c is less than 0.35. Also none of the m-derived sections with m = 0.4 to 0.7 gives 30 dB attenuation with f/f_c = 0.5-0.6 therefore these are also rejected.

Then, considering m = 0.8 and 0.9 and the more stringent requirement of 30 dB at 1200 Hz,

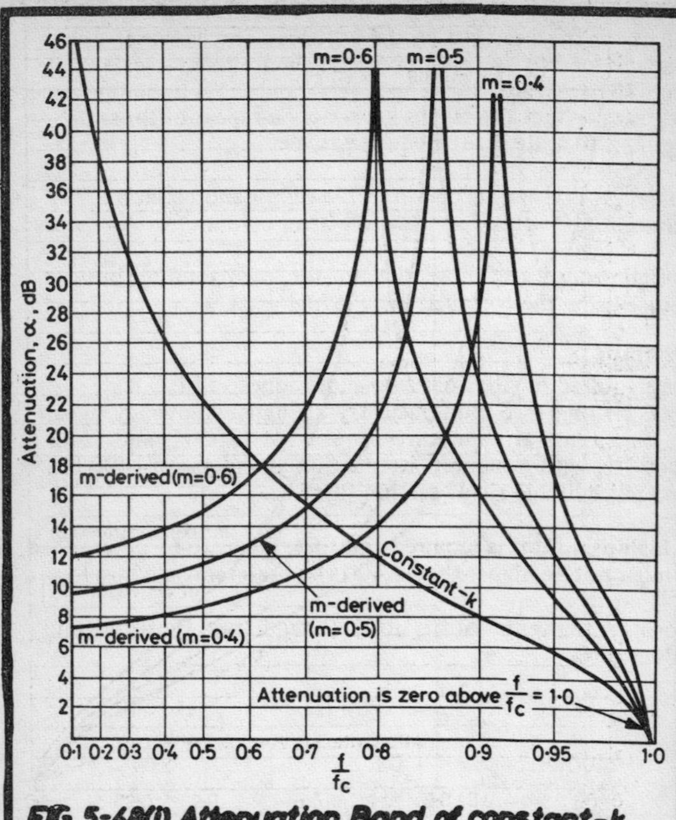

FIG. 5-48(i) Attenuation Band of constant-k and m-derived (m=0·4 -0·6) High-Pass Filters

with f_c = 2400 Hz, $\frac{f}{f_c}$ = 0.5 and m = 0.9 is satisfactory

" " = 2200 Hz " = 0.55 " = 0.8 is satisfactory

" " = 2000 Hz " = 0.6 " = 0.8 is satisfactory and in fact gives very high attenuation.

In each case the requirement at 1600 Hz is amply met.

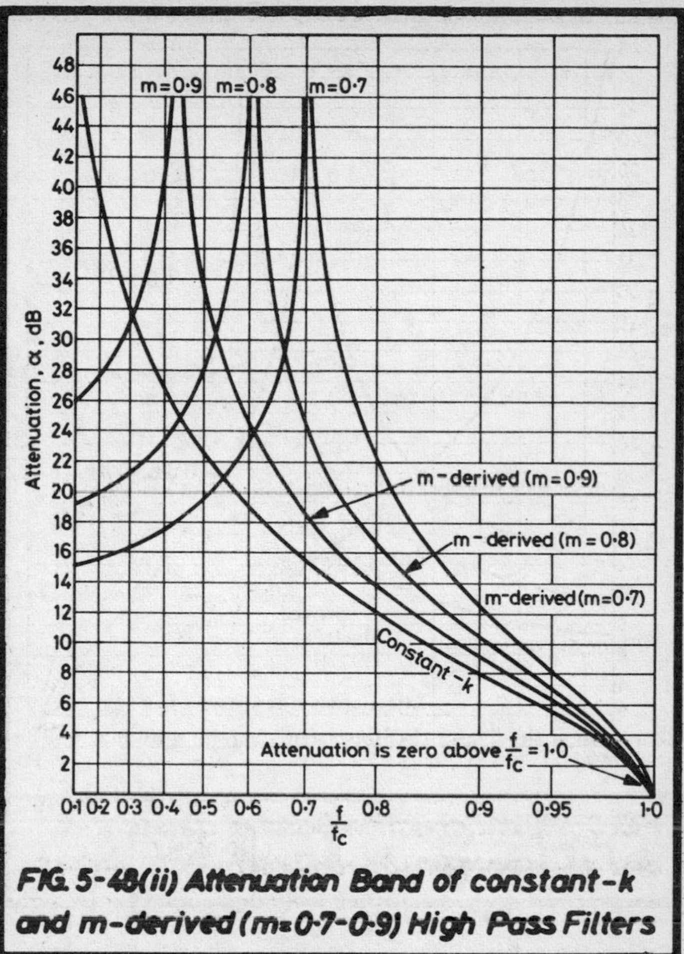

FIG. 5-48(ii) Attenuation Band of constant-k and m-derived (m=0·7–0·9) High Pass Filters

The best design would therefore appear to be for f_c = 2000 Hz with m = 0.8 because this gives plenty of room for manoeuvre in choice of component values and tolerances, but the other values for f_c = 2200 and 2400 Hz are technically suitable.

Summary of design requirements:

(i) $f_c = 2000$ Hz (ii) $R_o = 600$ ohms (iii) $m = 0.8$

Then:

$$L = \frac{600}{4\pi \times 2000} \text{ H} = 23.9 \text{ mH} \quad \frac{L}{m} = 29.9 \text{ mH}$$

$$\frac{2L}{m} = 59.8 \text{ mH}$$

$$C = \frac{10^6}{4\pi \times 2000 \times 600} \mu\text{F} = 0.0663 \ \mu\text{F} \quad \frac{C}{m} = 0.0829 \ \mu\text{F}$$

$$\frac{2C}{m} = 0.1658 \ \mu\text{F}$$

$$\frac{4m}{1-m^2} L = 8.889L = 0.212 \text{ H}$$

$$\frac{4m}{1-m^2} C = 8.889C = 0.589 \ \mu\text{F}$$

from which the T and Π filter circuits can be drawn as in Fig.5.49.

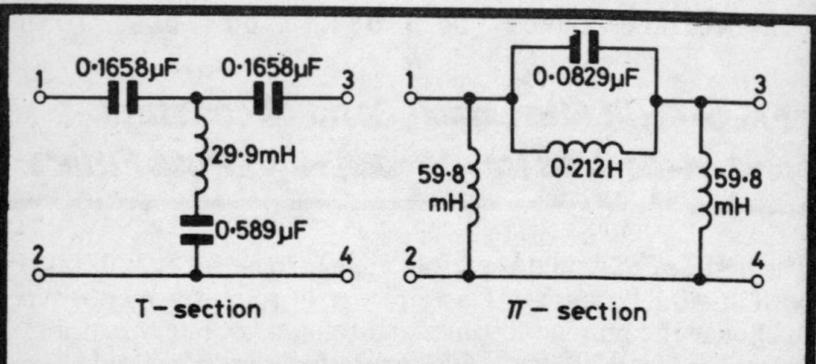

FIG. 5-49 m-derived High-Pass Filter Sections, f_c = 2000Hz, R_o = 600 ohms, m = 0·8

The practical filter:
Unless other considerations apply, the T-section is the obvious choice because it contains one inductor only compared with three in the π-section. The series capacitors may be 0.16 with 0.0056 μF in parallel, giving 0.1656 μF with practically no change in performance of the filter. The shunt arm components may be an inductor of 30 mH (a slight increase) with a capacitor 0.56 in parallel with 0.027 μF giving a slight decrease, both changes resulting in practically no change in f_r. The filter will therefore meet the specification with attenuation to spare.

6. MEASUREMENTS

6.1 EXTENSION OF THE RANGE OF MOVING-COIL INSTRUMENTS

Ammeters for general electronic construction work are available at full scale deflexion currents (f.s.d.) upwards from 50 μA or even less. The more sensitive the movement is (lower f.s.d. current), the greater the ohms-per-volt value will be when arranged for the measurement of voltage. A high ohms-per-volt value is important so that the voltmeter does not appreciably affect the circuit it is measuring. Equally a low f.s.d. current means that smaller currents will be indicated by larger movements of the needle and thus measured more accurately but in this case the movement resistance must not be forgotten because as an ammeter, the movement on its own will add an appreciable resistance to the circuit.

Many movements are marked with both the f.s.d. current and the movement resistance. However, all too frequently, the latter is omitted and this must therefore first be measured.

6.1.1 Measurement of Movement Resistance

There are obviously several ways of measuring the movement resistance but care must be taken that current greatly in excess of the f.s.d. value is not passed or the familiar expression of "wrapping the needle round" becomes a distinct possibility.

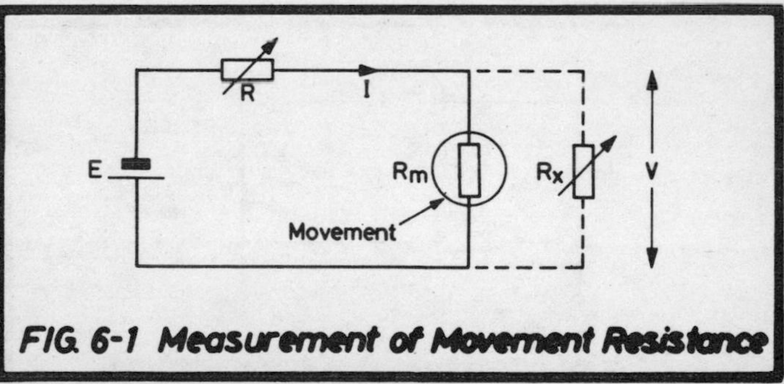

FIG. 6-1 Measurement of Movement Resistance

A simple method is by using a constant current supply as in Fig.6.1 and this method is suitable for all moving coil instruments because deflexion is proportional to current.

As an example, suppose the meter under test has an f.s.d. of 100 μA. The resistor R is chosen to be of such value that with the battery voltage used, f.s.d. current is obtained (as read on the meter). A reasonably high battery voltage is used (say, in this case, 6 V) so that R is high, e.g. neglecting R_m,

$$R = \frac{6}{100 \times 10^{-6}} = 60{,}000 \text{ ohms}$$

and it is obvious that a movement resistance of 100 ohms or so will affect the current only slightly. Power dissipated in R is very small, thus an ordinary radio potentiometer (e.g. 100 kΩ) or selection of fixed preferred value resistors will suffice. If a potentiometer is used, its value is reduced from maximum until f.s.d. is obtained. Next a variable low resistor R_x is connected across the meter and adjusted until the reading falls to half (or fixed resistors tried for the same result). Since the current supply I is constant, then equal currents now flow through the meter and R_x, hence because the voltage V across them both is the same,

$$R_m = R_x.$$

6.1.2 Ammeters

Let the currents and resistances be symbolized as in Fig.6.2.

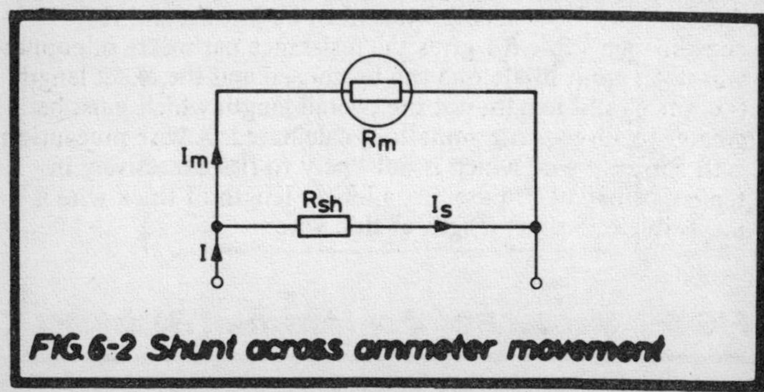

FIG.6-2 Shunt across ammeter movement

Then $I_s.R_{sh} = I_m R_m$

and $I = I_{sh} + I_m$

$$\therefore \text{ multiplying factor of shunt, } n = \frac{I}{I_m} = \frac{R_{sh} + R_m}{R_{sh}}$$

This is the factor by which the ammeter reading (with shunt) must be multiplied to give the true value of measured current, I.

$$\text{Also} \quad R_{sh} = \frac{R_m}{n - 1}$$

which enables the value of a shunt resistance to be calculated.

EXAMPLE:
Suppose a 5 mA movement of 20 ohms resistance is required to measure currents up to 500 mA. What value of shunt is required?

$$n = \frac{500}{5} = 100 \qquad R_{sh} = \frac{20}{100 - 1} = 0.202 \text{ ohms.}$$

This example can be used to illustrate the practical problem of obtaining a temporary meter shunt when a sufficiently precise resistance measuring instrument is not available, it is quite simply solved by use of an exact length of a resistance wire (see Section 3.1.2) or what is more likely to be at hand, a length of copper wire. Using the latter method, the gauges or diameters of the available wires must be known and for convenience Table 6.1 gives the resistance per metre of copper wires. An appropriate one can be chosen and the exact length (i.e. the circuit length, not the overall length which must be greater to allow for termination) calculated. A wise precaution is to choose a wire which is not likely to rise excessively in temperature when in use, i.e. a longer length of thick wire is preferable to a short length of thin wire.

Diameter mm	Resistance/metre ohms
2.6	.003247
2.4	.003811
2.2	.004535
2.0	.005488
1.8	.006775
1.6	.008575
1.4	.01120
1.2	.01524
1.1	.01814
1.0	.02195
0.9	.02698
0.8	.03430
0.7	.04480
0.6	.06098
0.55	.07257
0.5	.08781
0.45	0.1084
0.4	0.1372
0.35	0.1792
0.3	0.2439
0.28	0.2800
0.26	0.3247
0.24	0.3811
0.22	0.4535
0.20	0.5488
0.18	0.6775
0.16	0.8575
0.14	1.120
0.12	1.524
0.11	1.814
0.10	2.195
0.09	2.698
0.08	3.430

S.W.G.	Resistance/metre ohms
12	.003146
13	.004020
14	.005440
15	.006563
16	.008307
17	.01085
18	.01477
19	.02127
20	.02625
21	.03323
22	.04340
23	.05907
24	.07030
25	.08506
26	0.1050
27	0.1265
28	0.1553
29	0.1840
30	0.2213
31	0.2529
32	0.2917
33	0.3402
34	0.4020
35	0.4822
36	0.5891
37	0.7358
38	0.9451
39	1.258
40	1.477
41	1.757
42	2.127
43	2.625
44	3.323
45	4.340

Table 6.1 **Resistances of Copper Wires per Metre**

Continuing the example in which a 0.202 ohm shunt is required, the table shows for example, copper wire of SWG 30 has a resistance per metre of 0.2213 ohms, hence for 0.202 ohms:

$$\text{length required} = \frac{100}{0.2213} \times 0.202 \text{ cms} = 91.28 \text{ cms}$$

This can be cut easily to a fraction of 1% and it is not too inconvenient a length to be accommodated on the workbench.

6.1.3 Voltmeters

To convert an ammeter into a voltmeter, a series resistance R_{se} is required as in Fig.6.3

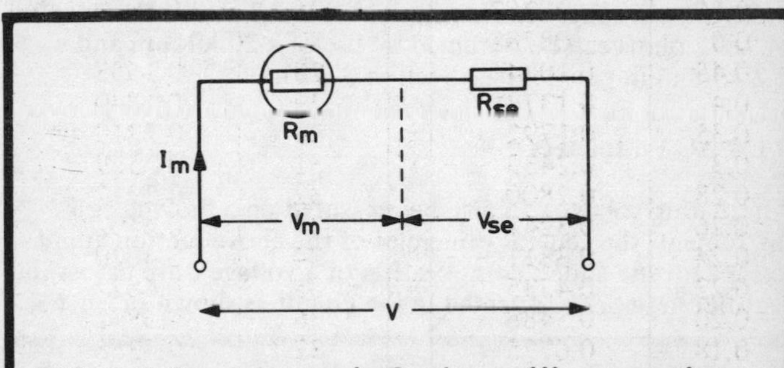

FIG. 6-3 Resistance in Series with ammeter movement for voltage measurement

$$V = V_m + V_{se}$$

$$V_m = I_m R_m$$

$$V_{se} = I_m R_{se}$$

Hence $$R_{se} = \frac{V}{I_m} - R_m$$

EXAMPLE:
Suppose a 5 mA movement with resistance 20 ohms is required to cover the voltage range 0-10 volts. What value of series resistor is required?

$$R_{se} = \frac{10}{.005} - 20 \text{ ohms} = 1980 \text{ ohms}$$

The total meter resistance is now 2000 ohms, therefore the meter has an ohms-per-volt value of 2000/10 = 200. In general the ohms-per-volt value (the higher this is, the more efficient is the voltmeter) is given by the reciprocal of the current for f.s.d. (in ampères), i.e. $1/I_m$, in the above example 1/.005 = 200.

The practical problem of obtaining a series resistor of such value is less than for the shunt case because a much lower current has to be carried and resistor values lie in the practical radio resistor range. For example, Table 3.7 shows that a value of 1892 ohms can be obtained by use of a 20 kilohm and a 2.2 kilohm in parallel (see Section 3.2.3).

6.1.4 AC Voltmeters

Alternating voltage can also be measured on a moving coil instrument, the general principles of the above section apply except for the added complication of a voltage drop across the rectifier network connected in the circuit as shown in Fig.6.4.

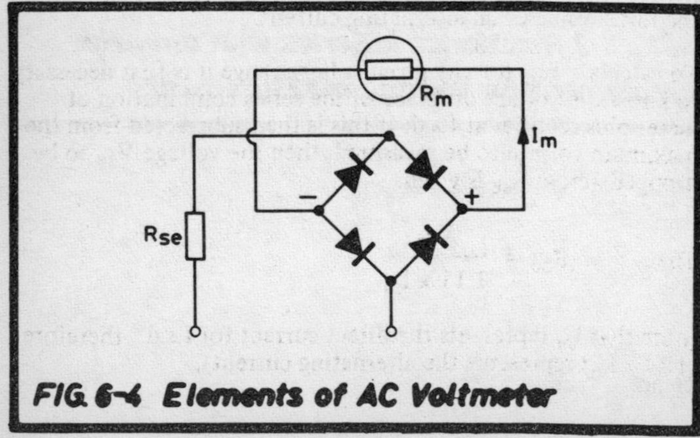

FIG. 6-4 Elements of AC Voltmeter

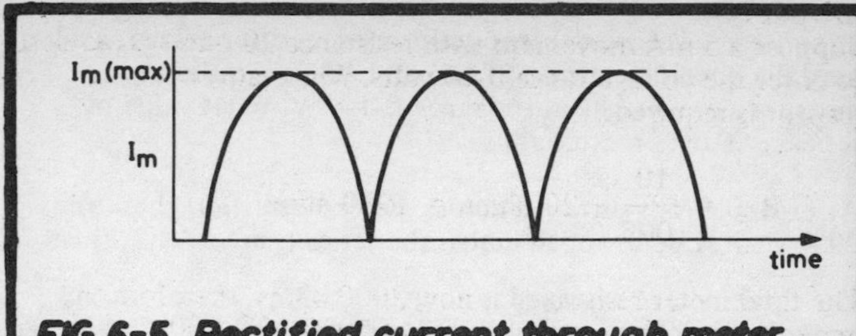

FIG.6-5 Rectified current through meter

A sine voltage applied to the full-wave rectifier bridge results in a current through the meter as in Fig.6.5. With perfect rectification, the meter deflexion will be proportional to the mean value. However it is normally required to measure the r.m.s. value.

Since mean value = 0.637 x max. value

and r.m.s. value = 0.707 x " "

$$\therefore \quad \frac{\text{r.m.s.}}{\text{mean}} = 1.11,$$

the familiar figure for the "form factor" of a sine wave. Thus the reading of a d.c. meter must be multiplied by 1.11 to give the r.m.s. value of an alternating current.

To calculate R_{se} for any given voltage range it is first necessary to know the voltage drop across the series combination of meter plus rectifier at f.s.d. If this is then subtracted from the maximum voltage to be measured, then the voltage V_{se} to be dropped across R_{se} is given.

Then $$R_{se} = \frac{V_{se}}{1.11 \times I_m}$$

(note that I_m represents the direct current for f.s.d., therefore $(1.11 \times I_m)$ represents the alternating current).

EXAMPLE:
Suppose a 1 mA movement with series rectifier bridge, having a total a.c. voltage drop at f.s.d. of 0.6 V is required to measure a.c. voltages in the range 0-100 V. What value of series resistor is required?

At f.s.d. 0.6 V is dropped across meter + rectifier, therefore 99.4 V must be dropped across the series resistor, R_{se}, then

$$R_{se} = \frac{99.4}{1.11 \times 1 \times 10^{-3}} = 89550 \text{ ohms.}$$

6.2 MEASUREMENT OF ELECTRICAL QUANTITIES

This section is intended to bring out the more salient points and basic formulae with regard to measurements which can be made in the amateur workshop or laboratory. The sophisticated engineer will possess multi-range meters, digital voltmeters, oscilloscope, counters, distortion analyser, and the like but much can be done with a handful of components, milliammeter, oscillator and a simple valve-voltmeter, which although now almost a relic of the past is a useful term used here to embrace all high-input-impedance voltmeters.

For bridge measurements at frequencies within the audio range, the valve-voltmeter can be replaced by the much less expensive earphone. Of the many methods available those shown in this section are probably the most practical and will produce results with sufficient accuracy for most purposes, but if however, greater accuracy is essential then certain precautions usually become inevitable and it is suggested that the more serious experimenter reads further from specialist text-books.

6.2.1 Resistance

(1) *Single meter methods*

The simplest method of measuring resistance is by the use of a battery and milliammeter as shown in Fig.6.6. Because of its simplicity, it is somewhat lacking in accuracy but is occasion-

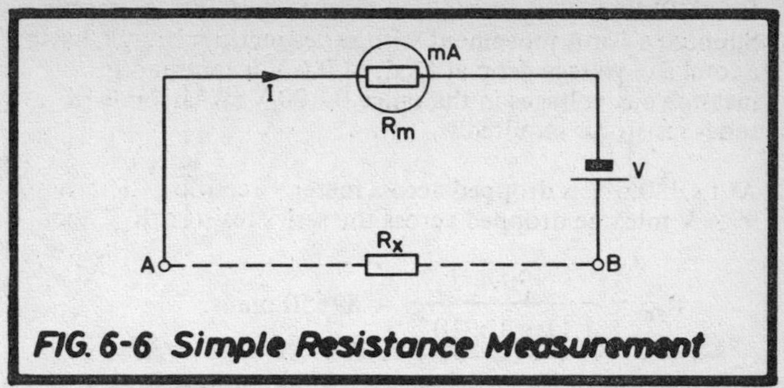

FIG. 6-6 Simple Resistance Measurement

ally useful when an approximate answer is acceptable. If the unknown resistance R_x is connected across A and B and the battery voltage (V) known (e.g. a 1.5 V cell), then the total circuit resistance R is given by Ohm's Law as V/I. However R includes both battery and meter resistances so these must be subtracted to give the true value for R_x. The meter resistance is probably known (or can be measured as shown in Sect.6.1.1) and an allowance of say 2-3 ohms for the single 1.5 V cell would be appropriate. This latter value varies with the condition of the battery and is the main source of inaccuracy although this is small if R_x is comparatively high. The method which follows does not suffer from this. The danger of damaging a sensitive meter if terminals A and B become short-circuited, or R_x is very low can be overcome by a resistor in series with the battery, this resistor being short-circuited when a reading is taken. Fig.6.7 shows a circuit in which compensation for battery internal resistance changes is made:

Terminals A and B are short-circuited and R is adjusted for full-scale deflexion (f.s.d.). A known resistance R_x is connected and the (lower) reading noted. Other values of R_x are then substituted to cover the whole scale which can either be marked directly with resistance values or alternatively a separate graph may be plotted relating resistance values to deflexion. Changes in battery resistance are accommodated by readjusting R as necessary. Again, a current limiting resistor may be an advantage.

The method of calibration in fact converts the milliammeter into an ohmmeter. For a single or a few measurements only,

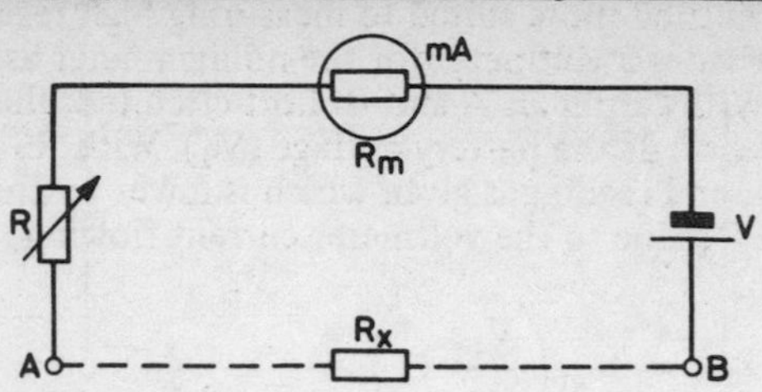

FIG. 6-7 Resistance Measurement with Compensation for Battery Resistance

full calibration is unnecessary and the following formula may be used provided that a particular value used for R is known, this value need not produce f.s.d. With terminals A and B short-circuited, the meter current is read (I_1), then with R_x in position, a second (lower) reading is obtained (I_2)

Then $$R_x = (R + R_m)\left(\frac{I_1}{I_2} - 1\right)$$

The above methods are not well suited to measurement of high resistances e.g. above 1 megohm and even this value would need a sensitive meter and relatively high battery

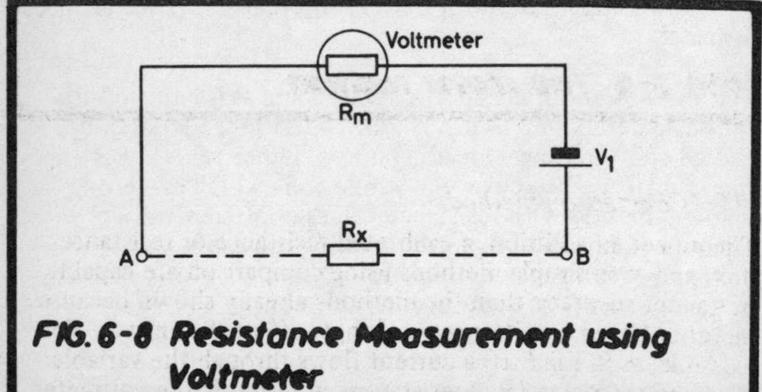

FIG. 6-8 Resistance Measurement using Voltmeter

voltage. A method more suited to measuring high resistance values substitutes a voltmeter for the milliammeter as shown in Fig.6.8. With terminals A and B short-circuited, the voltmeter simply reads the battery voltage (V_1). With R_x connected, a second reading is given which is lower by the voltage drop across R_x due to the voltmeter current flowing, call this V_2.

Then $$R_x = R_m \left(\frac{V_1}{V_2} - 1 \right)$$

(2) *Two-meter methods*

If two measuring instruments are available then the measurement technique shown in Fig.6.9 is a straightforward Ohm's Law one i.e. a current is made to flow through the unknown resistance R_x, the voltage across R_x is also measured and the value of R_x calculated. Some inaccuracy arises from the current taken by the voltmeter (which adds to the reading on the milliammeter), this is greatly reduced if a valve-voltmeter is used. A single multi-purpose meter can of course be used in place of the two separate meters.

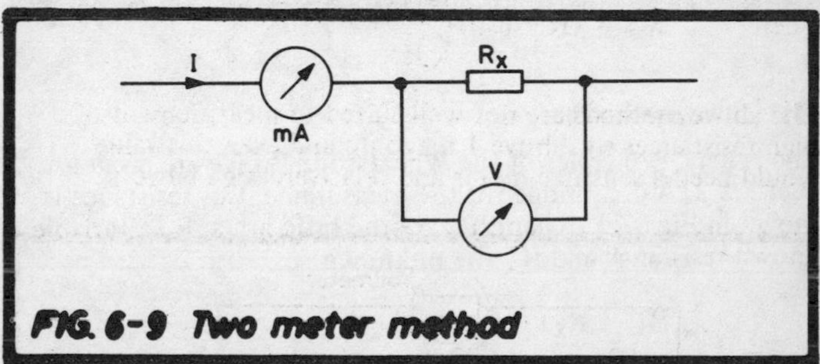

FIG. 6-9 Two meter method

(3) *Comparison methods*

These need in addition, a calibrated resistance or resistance box, and even simple methods using comparison are capable of greater accuracy than the methods already shown because meter or battery resistances no longer affect the answer. For example, as in Fig.6.10 a current flows through the variable (R_v) and unknown (R_x) resistances in series. If the voltmeter is switched between them and R_v adjusted for the same reading as obtained from R_X, then since the current I is the same through both resistances

$$R_X = R_V .$$

Equally a circuit could be set up in which measurements are made on R_X (i.e. voltage across, or current through) and R_V then used to replace R_X and adjusted for the previous result.

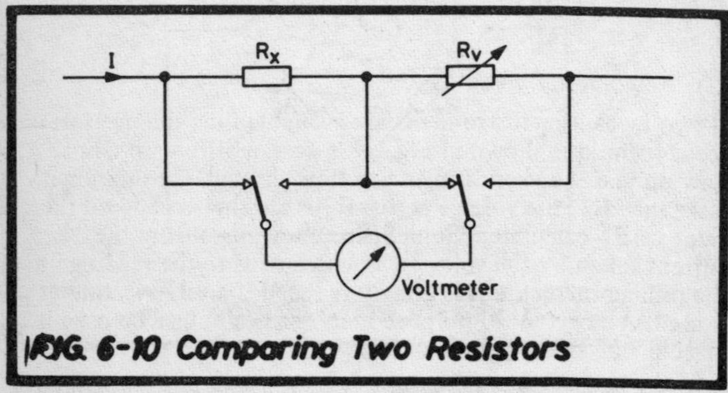

FIG. 6-10 Comparing Two Resistors

(4) *Wheatstone Bridge*

The general principles of the Wheatstone Bridge are given in Sect.5.1.2. A d.c. bridge for the measurement of resistance is shown in Fig.6.11. R_1 and R_2 are the ratio arms, R_V a variable known resistance and R_X the unknown.

Then $$\frac{R_1}{R_2} = \frac{R_X}{R_V}$$

and $$R_X = R_V \cdot \frac{R_1}{R_2}$$

When $R_1 = R_2$ the range of measurement is limited to the range of R_V. If $R_1 = 10$ or $100 \times R_2$ resistances much higher can be measured and similarly, if $R_2 = 10$ or $100 \times R_1$ resistances much lower can be measured. Thus the R_1/R_2 ratio can easily range over 10^4. Typically both may have a choice between three values, 10, 100 or 1000 ohms and if R_V is variable over say 1 to 10,000 ohms, the bridge will have a range of 10^8 i.e. from 0.01 ohms to 1 megohm.

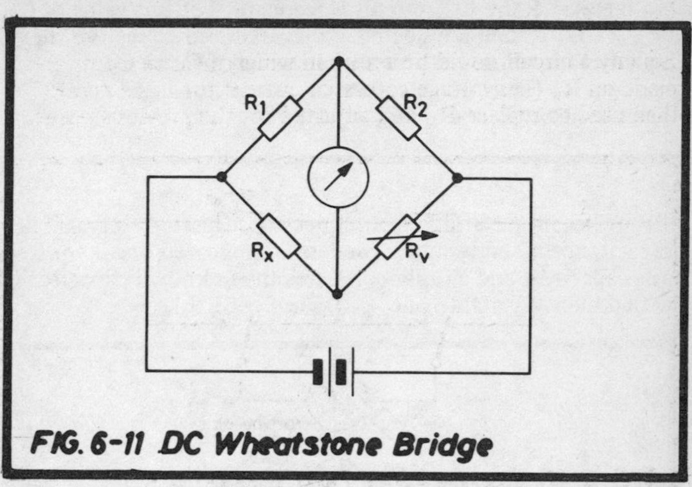

FIG. 6-11 DC Wheatstone Bridge

6.2.2 Capacitance

Small capacitors (e.g. in the picofarad range) may be measured by the resonance method shown in Fig.6.12.

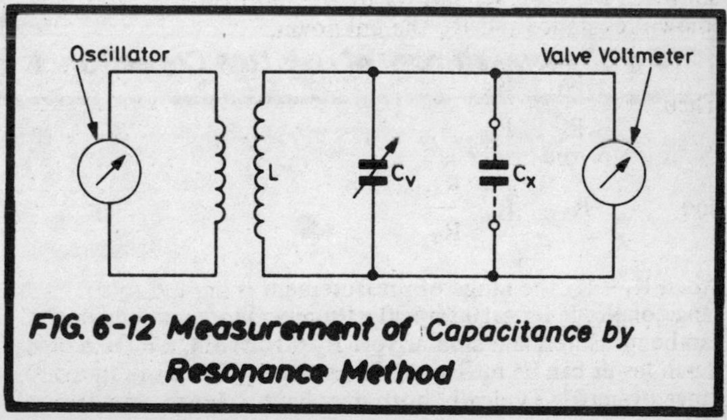

FIG. 6-12 Measurement of Capacitance by Resonance Method

C_V is a variable, calibrated capacitor and C_X the unknown. With C_X not connected, a convenient oscillator frequency is chosen and C_V is adjusted for maximum reading on the valve-voltmeter, i.e. the LC_V circuit is resonant. Let this value of C_V be C_{V_1} . C_X is then connected and C_V reduced to restore the maximum reading. Let this second value of C_V be C_{V_2}.

Then because $C_{V_1} = C_{V_2} + C_X$

$$C_X = C_{V_1} - C_{V_2}$$

Two Wheatstone Bridge techniques are suggested for general capacitance measurement. The less complicated one is for low loss capacitors and simply compares the unknown capacitor C_X with a known variable one, C_V, as in Fig.6.13.

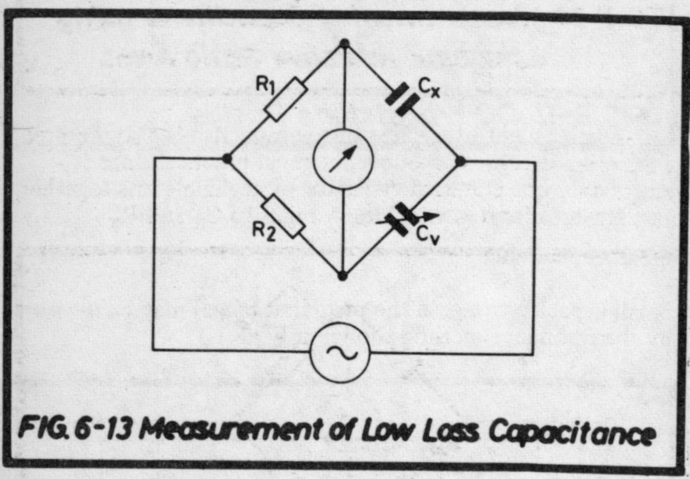

FIG. 6-13 Measurement of Low Loss Capacitance

Solution of the bridge is

$$C_X = C_V \cdot \frac{R_1}{R_2}$$

and balance is obtained by adjustment of C_V or more conveniently by using a fixed value for C_V and altering the ratio of R_1 to R_2 as in Fig.6.14. R_1 and R_2 together form a calibrated potentiometer.

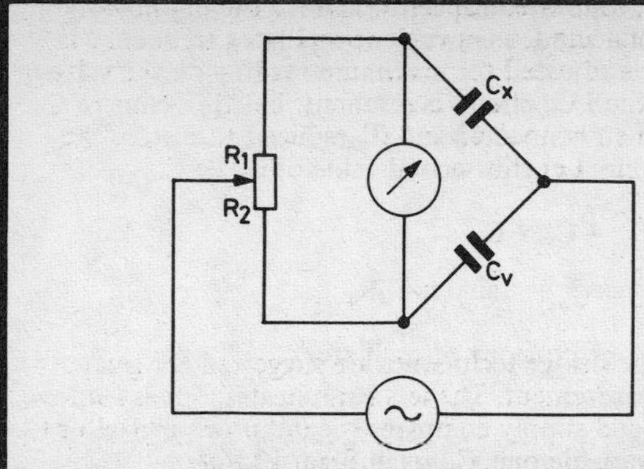

FIG. 6-14 Measurement of Capacitance using Adjustable Resistive Ratio Arms

When measurement of the loss of the capacitor is also required, the Schering Bridge is probably the most practical since it requires only one standard capacitor of negligible loss together with a small calibrated variable. In Fig.6.15 C_x and R_x

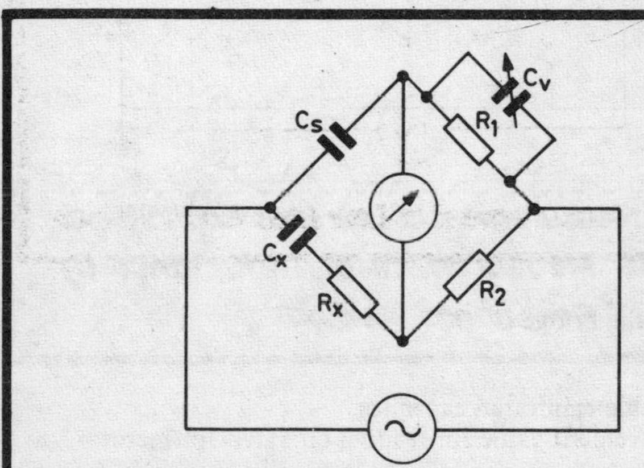

FIG. 6-15 Measurement of Capacitance with Loss

represent the unknown capacitor, C_s is the standard fixed capacitor and C_V the variable. Balance is given by

$$C_X = C_S \cdot \frac{R_1}{R_2} \qquad R_X = R_2 \cdot \frac{C_V}{C_s}$$

$$\tan \delta = \omega C_V R_1$$

$$\therefore \text{ loss angle } \delta = \tan^{-1} \omega C_V R_1$$

(where $\omega = 2\pi \times$ frequency of oscillator).

6.2.3 Inductance, Mutual Inductance and Q-Factor

As with capacitance, inductance can also be measured by a resonance method but in the circuit shown (Fig.6.16) the self-capacitance of the coil and the small effect of the coupling coil are neglected.

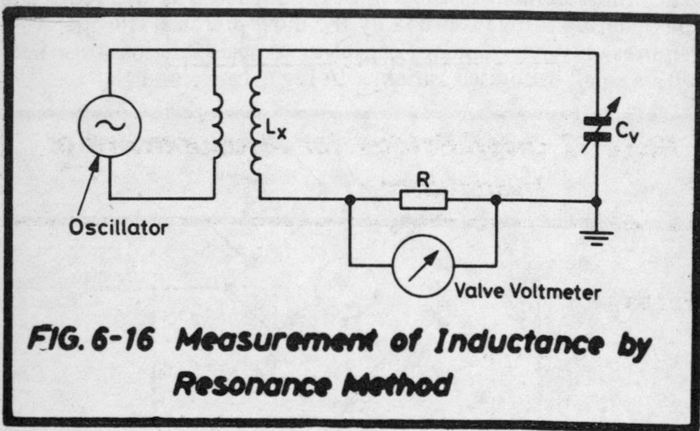

FIG. 6-16 Measurement of Inductance by Resonance Method

C_V is a variable calibrated capacitor
R is of convenient value for reading on valve-voltmeter

Then at frequency f, C_V is tuned for maximum reading and

$$L_X = \frac{1}{4\pi^2 f^2 C}$$

There are several Wheatstone Bridge circuits available for measurement of inductance e.g. Maxwell's method of balancing an unknown inductance with a known one or an alternative method of balancing an unknown inductance with a capacitance. The latter is more practical since a variable standard inductor is less likely to be available than a variable capacitor. However, the Owen Bridge is better still because it can employ fixed capacitors with variable resistance. Fig.6.17 refers.

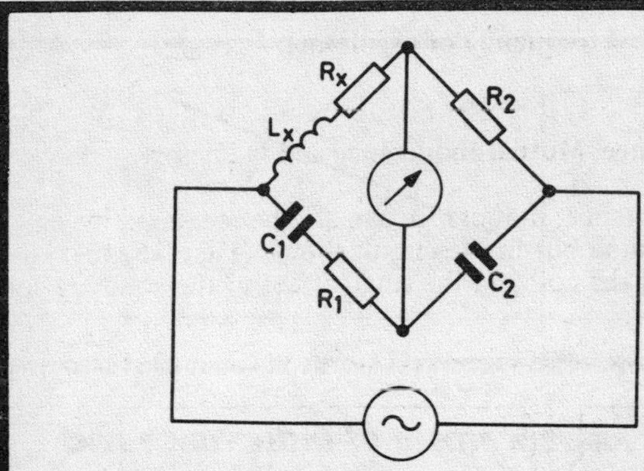

FIG.6-17 Owen Bridge for Measurement of Inductance

For balance

$$L_x = C_2 R_1 R_2$$

$$R_x = R_2 \cdot \frac{C_2}{C_1}$$

Thus usually C_1 and C_2 are fixed and balance is obtained by adjustment of R_1 and a resistance box in series with L_x and R_x, the known value of the added resistance then being subtracted from the value of R_x given above to obtain the true resistive component of the inductor.

Neither of the above two methods produces a value for the self-capacitance (C_S) of an inductor. This may be obtained by a graphical method using the test circuit already given in Fig.6.16. The values of C_V for resonance (maximum reading) over a range of frequencies are noted. Then plot C_V against $1/f^2$ as in Fig.6.18. This should give a straight-line graph which can be extended back to cut the C_V axis on the apparently negative side and it is this intercept which gives the value of C_S as shown. C_S is the lumped self-capacitance of the whole circuit but most of it is associated with the inductor.

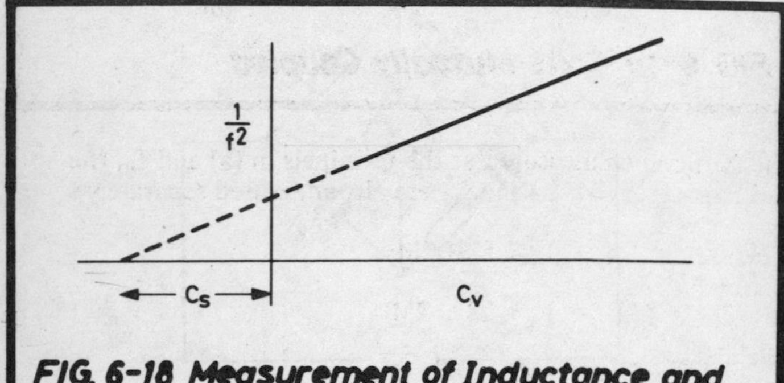

FIG. 6-18 Measurement of Inductance and Self-Capacitance by Resonance Method

The inductance is given by the slope of the line since

$$\text{Slope} = \frac{\frac{1}{f^2}}{C_V} = \frac{1}{f^2 C_V} \quad \text{and since} \quad L_X = \frac{1}{4\pi^2 f^2 C_V}$$

$$\text{Slope} = 4\pi^2 L_X.$$

Mutual Inductance

If an inductance measuring bridge is available, mutual inductance can also be measured. Consider the two mutually coupled coils L_1 and L_2 connected in series aiding in Fig.6.19(a) and in series opposition in (b) [the dots indicate that when the current enters (or leaves) both coils at these ends the mutual inductance is +ve and vice versa]. Let L_A be

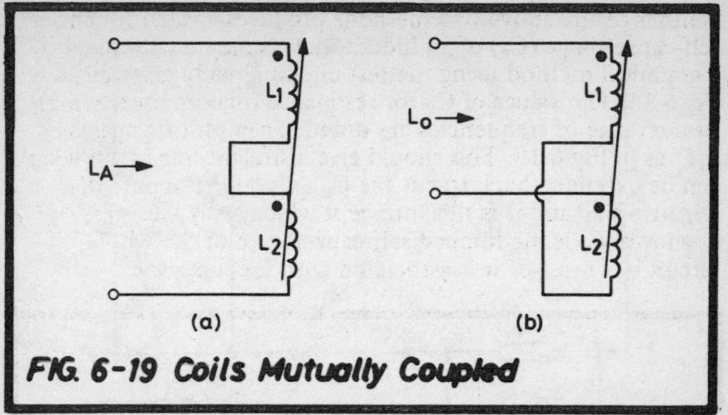

FIG. 6-19 Coils Mutually Coupled

the inductance measured at the terminals in (a) and L_o the inductance in (b). L_1 and L_2 are also measured separately.

$$\text{Then} \qquad L_A = L_1 + L_2 + 2M$$

$$L_o = L_1 + L_2 - 2M$$

$$L_A - L_o = 4M$$

$$\text{and} \qquad M = \frac{L_A - L_o}{4} \quad \text{(all values will be complex).}$$

Q-Factor

This is usually required at a resonant frequency and a circuit for measurement is given in Fig.6.20.

$L_x + R_x$ is the inductor under test.

R_1 is a small non-reactive resistor. The valve voltmeter across R_2 measures the oscillator current I which flows through R_1 and therefore injects a voltage IR_1 into the tuned circuit. C_v is adjusted for resonance as indicated with the valve voltmeter connected across it as shown. The valve voltmeter reading is noted, V.

$$\text{Then} \quad Q = \frac{V}{IR_1}$$

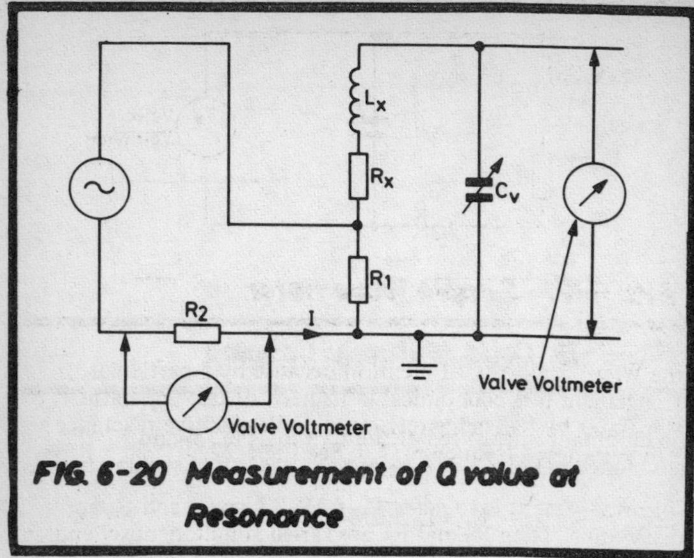

FIG. 6-20 Measurement of Q value at Resonance

6.2.4 Frequency

A simple tuned frequency meter, more commonly known as a wavemeter as shown in Fig.6.21 uses a pick-up coil as its inductance and also for coupling to aerial, oscillator or other circuit carrying the frequency to be measured (f_x). C is adjusted for resonance and if L is known and C is calibrated, then provided that the resistive component (R) of L is reasonably small compared with ωL,

$$f_x = \frac{1}{2\pi\sqrt{LC}}$$

Frequently the wavemeter is calibrated directly in frequency on the dial of C so that no calculation is involved.

Frequency measurements are most conveniently made by use of a calibrated time base of an oscilloscope, but one bridge

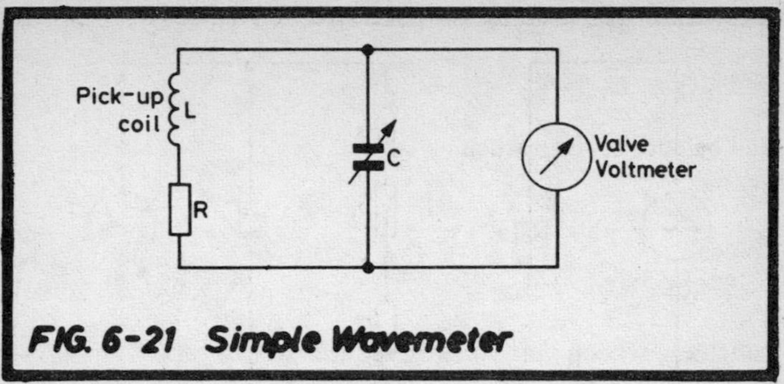

FIG. 6-21 Simple Wavemeter

(the Wien) is worthy of mention because by a particular arrangement it is economical in its need of test apparatus. It operates by balancing resistance and capacitive reactance at the unknown frequency (f_x). See Fig.6.22.

The arrangement is to make $R_4 = 2R_3$. Also C_1 and C_2 are made equal. Then R_1 and R_2 are varied simultaneously and also kept equal.

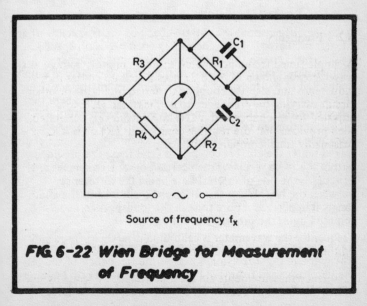

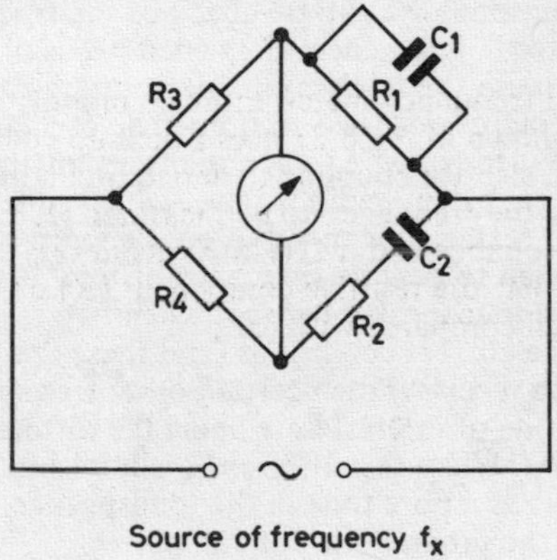

Source of frequency f_x

FIG. 6-22 Wien Bridge for Measurement of Frequency

Thus let $R_1 = R_2 = R$

$C_1 = C_2 = C$

and balance is given when

$$\frac{2R_3}{\frac{1}{R} + j\omega C} = R_3\left(R + \frac{1}{j\omega C}\right)$$

which expression conveniently reduces to

$$\omega^2 = \frac{1}{C^2 R^2}$$

and therefore $f_x = \dfrac{1}{2\pi CR}$

The frequency range of the bridge can be altered by changing C_1 and C_2 to other fixed and equal values.

Check of RF Test Oscillators:—

A simple but reasonably accurate check of the frequency of an r.f. test oscillator (signal generator) which needs no special equipment is to use the freely available accurate carrier frequencies of radio broadcast stations. The oscillator is simply connected to a piece of wire to act as a transmitting aerial and this is placed near to any radio receiver which tunes over the range required. A station is tuned in on the radio and the oscillator adjusted to near its frequency. This will produce a beat-note in the loudspeaker because when two frequencies are mixed under the correct conditions (and these are provided within the radio receiver) then certain other frequencies are produced, the one of interest here being the difference between the two. When this difference falls within the audio range, it appears as a pure tone in the loudspeaker, superimposed upon the programme but distinctive.

Take, for example, comparison with a station on 1214 kHz. The graph in Fig. 6.23 shows that as the frequency of the test oscillator is varied from 1211 to 1217 kHz a note of 3 kHz → 0 → 3 kHz will appear. Between about 123.9 and 124.1 kHz, the note is within the range 0-100 Hz and this possibly will not be reproduced by the loudspeaker, even if it were, there is certainly a small range or 'dead space' centering on 1214 kHz. The oscillator should be adjusted by 'feel' to the centre of this space. In any case, although the 'dead space' may be as much as 200 Hz wide, this on the average r.f. test oscillator is frequently hardly a discernible movement of the pointer. Thus the test oscillator scale can be checked or calibrated at this point and other radio transmitters subsequently tuned in for further checks to be made.

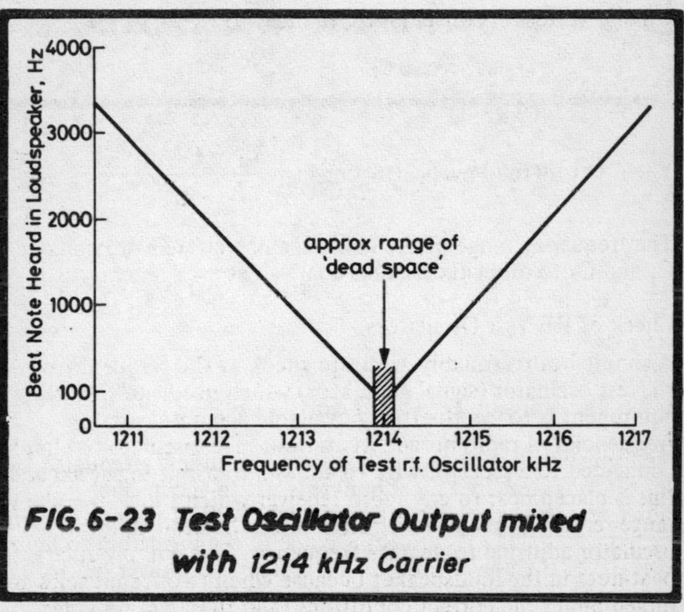

FIG. 6-23 Test Oscillator Output mixed with 1214 kHz Carrier

6.2.5 Power

Measurement of power is not often required in the amateur workshop or laboratory but a simple method suitable for low power mains or higher frequency circuits is worth recording because the only measuring device it needs is a high resistance voltmeter (or valve-voltmeter). Three voltage measurements are made across and each side of a single resistance, R, as shown in Fig.6.24.

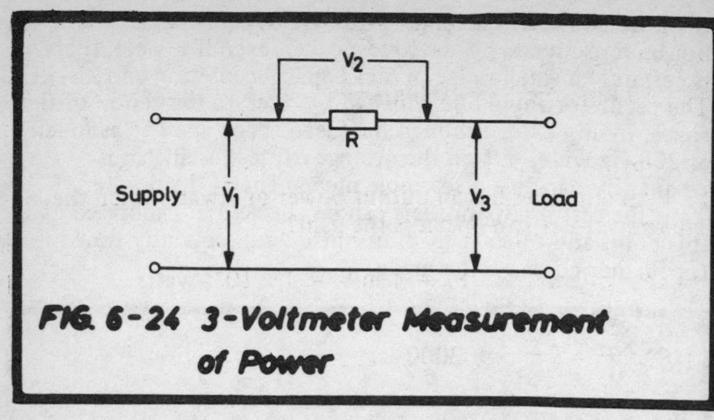

FIG. 6-24 3-Voltmeter Measurement of Power

Then $\quad$ True Power into Load $= \dfrac{V_1^2 - (V_2^2 + V_3^2)}{2R}$

Power Factor $(\cos \phi) = \dfrac{V_1^2 - (V_2^2 + V_3^2)}{2V_2 V_3}$

6.2.6 Transmission Loss and Gain

Transmission loss or gain in a system (amplifier, attenuator, line etc.) is measured most conveniently in the decibel notation. The basic unit is actually the Bel and the number of bels transmission loss or gain is equal to the logarithm to the base 10 of the power ratio of the system (i.e. power out to power in). The decibel is one-tenth of the bel and is a unit of more convenient magnitude for general work. It is defined as follows:

If P_1 and P_2 are the input and output powers respectively of a system,

$$\text{No. of decibels} = 10 \log_{10} \frac{P_2}{P_1}$$

Although the numerical answer is the same whether P_2/P_1 or P_1/P_2 is considered, a positive sign is used to express gain in a system, and a negative sign, loss. Note that

$$10 \log_{10} \frac{P_2}{P_1} = -10 \log_{10} \frac{P_1}{P_2}$$

EXAMPLE:
(1) An amplifier has an output power of 2 watts when the input power is 1 mW. What is the gain?

$$P_2 = 2 \text{ watts}, \quad P_1 = 1 \text{ mW} = 1 \times 10^{-3} \text{ watts}$$

$$\therefore \frac{P_2}{P_1} = \frac{2}{10^{-3}} = 2000$$

$$\log_{10} 2000 = 3.3010$$

$$\therefore \text{Gain} = 10 \times 3.3010 = 33.01 \text{ dB.}$$

(2) An attenuator delivers 1 mW of power when its input is 2 watts. What is the loss?

$$P_2 = 1 \text{ mW}, \quad P_1 = 2 \text{ watts}$$

$$\therefore \frac{P_2}{P_1} = \frac{1 \times 10^{-3}}{2} = 5 \times 10^{-4}$$

$$\log_{10} 5 \times 10^{-4} = \bar{4}.6990 = -4 + 0.6990 = -3.3010$$

$$\therefore \text{Loss} = 10 \times -3.3010 = -33.01 \text{ dB}$$

(the minus sign indicating that it is a loss).

Consider Table 6.2.

Input/Output Power Ratio	Bels	Decibels
0.001	−3	−30
0.01	−2	−20
0.1	−1	−10
1	0	0
10	1	10
100	2	20
1000	3	30

Table 6.2 **Bel and Decibel Equivalents of Power Ratios**

It is clear that when power ratio multiples of 10 are considered, the number of decimal places or noughts gives the loss or gain in bels. In decibels it is therefore ten times this figure.

Also, very nearly

A power ratio of 1.25 : 1 is equivalent to 1 dB
2 : 1 " " " 3 dB

and from the last figure it is seen that doubling or halving the power adds or subtracts 3 dB each time e.g. 4 x power = 3+3 = 6 dB. This provides a quick means of obtaining the approximate power ratio from a knowledge of the number of decibels.

EXAMPLE:
What are the approximate power ratios corresponding to 23 dB and 27 dB

23 dB = 20 dB + 3 dB

Corresponding power ratios to 20 dB and 3 dB are 100 and 2

∴ 23 dB represents a power ratio of 100 x 2 = 200

27 dB = 20 dB + 3 dB + 3dB + 1 dB

Corresponding power ratios are 100, 2, 2, 1.25 (or $\frac{5}{4}$)

∴ 27 dB represents a power ratio of $100 \times 2 \times 2 \times \frac{5}{4} = 500$

The precise answer to the latter is 501.2 showing that the quick method suffers very little from loss of accuracy.

The Neper system which is also in use, but mainly for theoretical work expresses current or voltage ratios logarithmically, but this time in natural logarithms (i.e. to the base e) thus:

$$\text{No. of Nepers} = \log_e \frac{I_1}{I_2}$$

and the decineper is equal to $\frac{1}{10}$ Neper

$$\text{so that No. of decinepers} = 10 \log_e \frac{I_1}{I_2}$$

It can be shown that 1 Neper is equivalent to 8.686 decibels
or 1 decibel " " " 0.1151 nepers

provided that nepers are measured in the same resistive components of the input and output of the system.

No mention has yet been made (except above with regard to nepers) of measuring transmission loss or gain by comparing input and output voltages or currents. Strictly this can only be done when input and output impedances are equal as developed below:

Consider a gain calculation where input and output voltages (V_1 and V_2) and impedances (Z_1 and Z_2) are known,

$$\text{Input power} = \frac{V_1^2}{Z_1} \qquad \text{Output power} = \frac{V_2^2}{Z_2}$$

$$\therefore \text{ Gain in decibels} = 10 \log_{10} \frac{V_2^2}{Z_2} \times \frac{Z_1}{V_1^2}$$

If $Z_1 = Z_2$

$$\text{Gain in decibels} = 10 \log_{10} \left(\frac{V_2}{V_1}\right)^2 = 20 \log_{10} \frac{V_2}{V_1}$$

and similarly for currents I_1 and I_2

$$\text{Gain in decibels} = 20 \log_{10} \frac{I_2}{I_1}$$

But this cannot be true unless $Z_1 = Z_2$. In spite of this current and voltage gains (especially the latter) are frequently erroneously quoted as $20 \log_{10}(V_2/V_1)$ irrespective of input and output impedance differences which, for example, in the case of an audio power amplifier may be very great. Thus an "amplifier voltage gain of 40 dB" is meant to imply that

$$40 = 20 \log_{10} \frac{V_2}{V_1}$$

$$\therefore \frac{V_2}{V_1} = \text{antilog } 2 = 100$$

Therefore, as the appropriate British Standard says, "to avoid confusion such use of the word 'decibel' should be accompanied by a specific statement giving in the particular case the quantities concerned."

However, what is legitimate is the practice of quoting decibel changes to show improvement or degradation at the same point in a system due to some action taken. In most cases the impedance is the same for both measurements and therefore 20 times the logarithm of the voltage or current ratio is valid. An example will make this clear.

EXAMPLE:
The mains hum voltage measured across the loudspeaker of an audio system is reduced from 0.2 mV to 0.1 mV. What is the decibel improvement?

Signs can nearly always be ignored in this type of case because it is obvious whether a change is an improvement or not.

Thus

$$\text{decibel improvement} = 20 \log_{10} \frac{0.2}{0.1} = 6.02$$

Alternatively it could be stated that the hum after reduction is −6 dB relative to its previous level.

For accurate assessment care must be taken to ensure that the impedance does not change from one measurement to the next – it may change with frequency, with signal level and even the output impedance of a transistor amplifier may be affected by changes within its earlier stages.

Standard Reference Levels:

Absolute levels can only be expressed in decibel notation by reference to a known, fixed quantity e.g. in line transmission the generally accepted reference is 1 mW and the impedance 600 ohms, non-reactive. To avoid having to quote this reference with each measurement the term dBm is used. It is for this reference level only and no other. Hence +20 dBm means 20 dB above the reference level, i.e. an absolute power level of 100 mW.

Another convenient reference level when working in terms of voltage is 1 volt in which case values are expressed in dB/1V.

A reference level being brought into greater use as aircraft and other noises increase is the one used in audio acoustics. This is the reference sound pressure level (SPL) of 2×10^{-5} Newtons per square metre (approximately 1 picowatt) which is the minimum sound pressure discernible by the human ear (measured at 1000 Hz on persons with normal hearing). Because sound pressure and electrical voltage have a direct relationship, sound pressure levels are compared by

$$20 \log_{10} \frac{P_1}{P_2} \text{ dB}$$ where P_1 and P_2 are the two levels being compared

Any sound level (P) is then quoted as SPL relative to 2×10^{-5} N/m^2 i.e.

$$\text{SPL} = 20 \log_{10} \frac{P}{2 \times 10^{-5}} \text{ dB.}$$

Noise creates many measurement problems compared with pure tones and modern "noise meters" incorporate special filters or "weighting networks" to resemble the listening acuity of the human ear over the audio range. The most commonly used network is referred to as the 'A' and this gives rise to noise sound pressure levels quoted as dBA. As an example, an office having a noise level of 60 dBA would produce this measurement on a noise meter using an A network, the level in fact being 60 dB (1000 times) above reference level, i.e. .02 N/m^2 sound pressure.

Other noise weighting networks are in use, including a special one for aircraft noise, the D network, giving rise to the unit dBD.

Measurement

Power, of course can be measured from the principles of Ohm's Law, i.e. voltages, currents, resistances/impedances (the latter are required if differing between input and output) but generally a system gain or loss is measured at any given frequency by connexion of an oscillator to the input and measurement at the output by a decibelmeter. The latter is an a.c. voltmeter, effective over the frequency range required and calibrated directly in decibels against the reference level of the oscillator.

APPENDIX

MARKING CODES FOR RESISTORS AND CAPACITORS

A system developed comparatively recently which is gaining in popularity uses a slightly abbreviated printed code. A single letter indicating the multiplier is added to the first two figures of the resistance or capacitance value and its position relative to these figures places the demical point which is therefore not shown. Only three multiples are needed, R for x 1, K for x 100 and M for x 1,000,000, conveniently remembered since K (but lower case) already stands for kilo (= 1000) and M for mega (= 1,000,000) in the normal decimal system. The examples which follow show the use of the multiples and demonstrate the method. For values less than 1 a 0 is used in the leading position.

The code	OR3	=	0.3 x 1	=	0.3
	2R7	=	2.7 x 1	=	2.7
	27R	=	27 x 1	=	27
	K27	=	0.27 x 1000	=	270
	2K7	=	2.7 x 1000	=	2,700
	27K	=	27 x 1000	=	27,000
	M27	=	0.27 x 1,000,000	=	270,000
	2M7	=	2.7 x 1,000,000	=	2,700,000
	27M	=	27 x 1,000,000	=	27,000,000

A fourth letter indicates the tolerance:

F = 1%, G = 2%, H = 2.5%, J = 5%, K = 10%, M = 20% (all ±).

thus a resistor marked 6K8K has a value of 6800 ±10% ohms.

A further method simply uses three digits followed by the tolerance letter (as above). The first two digits are those of the actual value while the third indicates the number of 0's which follow, thus a marking 682K represented 6800 ± 10% (ohms or say, picofarads).

Notes

Notes

Notes

Notes

Please note overleaf is a list of other titles that are available in our range of Radio, Electronics and Computer Books.

These should be available from all good Booksellers, Radio Component Dealers and Mail Order Companies.

However, should you experience difficulty in obtaining any title in your area, then please write directly to the publisher enclosing payment to cover the cost of the book plus adequate postage.

If you would like a complete catalogue of our entire range of Radio, Electronics and Computer Books then please send a Stamped Addressed Envelope to:–

BERNARD BABANI (publishing) LTD
THE GRAMPIANS
SHEPHERDS BUSH ROAD,
LONDON W6 7NF
ENGLAND

160	Coil Design and Construction Manual	£2.50
227	Beginners Guide to Building Electronic Projects	£1.95
BP28	Resistor Selection Handbook	£0.60
BP36	50 Circuits Using Germanium Silicon & Zener Diodes	£1.95
BP37	50 Projects Using Relays, SCRs and TRIACs	£2.95
BP39	50 (FET) Field Effect Transistor Projects	£2.95
BP42	50 Simple LED Circuits	£1.95
BP44	IC 555 Projects	£2.95
BP45	Projects in Opto-Electronics	£2.50
BP48	Electronic Projects for Beginners	£1.95
BP49	Popular Electronic Projects	£2.50
BP53	Practical Electronics Calculations & Formulae	£3.95
BP56	Electronic Security Devices	£2.95
BP58	50 Circuits Using 7400 Series IC's	£2.50
BP74	Electronic Music Projects	£2.95
BP76	Power Supply Projects	£2.50
BP78	Practical Computer Experiments	£1.75
BP80	Popular Electronic Circuits – Book 1	£2.95
BP84	Digital IC Projects	£1.95
BP85	International Transistor Equivalents Guide	£3.95
BP87	50 Simple LED Circuits – Book 2	£1.95
BP88	How to Use Op-amps	£2.95
BP90	Audio Projects	£2.50
BP92	Electronics Simplified – Crystal Set Construction	£1.75
BP94	Electronic Projects for Cars and Boats	£1.95
BP95	Model Railway Projects	£2.95
BP97	IC Projects for Beginners	£1.95
BP98	Popular Electronic Circuits – Book 2	£2.95
BP99	Mini-matrix Board Projects	£2.50
BP104	Electronic Science Projects	£2.95
BP105	Aerial Projects	£2.50
BP107	30 Solderless Breadboard Projects – Book 1	£2.95
BP110	How to Get Your Electronic Projects Working	£2.95
BP111	Audio	£3.95
BP112	A Z-80 Workshop Manual	£3.95
BP113	30 Solderless Breadboard Projects – Book 2	£2.25
BP115	The Pre-computer Book	£1.95
BP118	Practical Electronic Building Blocks – Book 2	£1.95
BP121	How to Design and Make Your Own PCB's	£2.50
BP122	Audio Amplifier Construction	£2.95
BP125	25 Simple amateur band Aerials	£1.95
BP126	BASIC & PASCAL in Parallel	£1.50
BP130	Micro Interfacing Circuits – Book 1	£2.75
BP131	Micro Interfacing Circuits – Book 2	£2.75
BP132	25 Simple SW Broadcast Band Aerials	£1.95
BP136	25 Simple Indoor and Window Aerials	£1.75
BP137	BASIC & FORTRAN in Parallel	£1.95
BP138	BASIC & FORTH in Parallel	£1.95
BP144	Further Practical Electronics Calculations & Formulae	£4.95
BP145	25 Simple Tropical and MW Band Aerials	£1.75
BP146	The Pre-BASIC Book	£2.95
BP147	An Introduction to 6502 Machine Code	£2.95
BP148	Computer Terminology Explained	£1.95
BP152	An Introduction to Z80 Machine Code	£2.95
BP171	Easy Add-on Projects for Amstrad CPC 464, 664, 6128 & MSX Computers	£2.95
BP174	More Advanced Electronic Music Projects	£2.95
BP176	A TV-DXers Handbook (Revised Edition)	£5.95
BP177	An Introduction to Computer Communications	£2.95
BP179	Electronic Circuits for the Computer Control of Robots	£2.95
BP180	Electronic Circuits for the Computer Control of Model Railways	£2.95
BP182	MIDI Projects	£2.95
BP184	An Introduction to 68000 Assembly Language	£2.95
BP187	A Practical Reference Guide to Word Processing on the Amstrad PCW8256 & PCW8512	£5.95
BP190	More Advanced Electronic Security Projects	£2.95
BP192	More Advanced Power Supply Projects	£2.95
BP193	LOGO for Beginners	£2.95
BP194	Modern Opto Device Projects	£2.95
BP195	An Introduction to Satellite Television	£5.95
BP196	BASIC & LOGO in Parallel	£2.95
BP197	An Introduction to the Amstrad PC's	£5.95
BP198	An Introduction to Antenna Theory	£2.95
BP230	A Concise Introduction to GEM	£2.95
BP232	A Concise Introduction to MS-DOS	£2.95
BP233	Electronic Hobbyists Handbook	£4.95
BP239	Getting the Most From Your Multimeter	£2.95
BP240	Remote Control Handbook	£3.95
BP243	BBC BASIC86 on the Amstrad PC's & IBM Compatibles – Book 1: Language	£3.95
BP244	BBC BASIC86 on the Amstrad PC's & IBM Compatibles – Book 2: Graphics and Disk Files	£3.95
BP245	Digital Audio Projects	£2.95
BP246	Musical Applications of the Atari ST's	£5.95
BP247	More Advanced MIDI Projects	£2.95
BP248	Test Equipment Construction	£2.95
BP249	More Advanced Test Equipment Construction	£3.50

BP250	Programming in FORTRAN 77	£4.95
BP251	Computer Hobbyists Handbook	£5.95
BP254	From Atoms to Amperes	£3.50
BP255	International Radio Stations Guide (Revised 1991/92 Edition)	£5.95
BP256	An Introduction to Loudspeakers & Enclosure Design	£2.95
BP257	An Introduction to Amateur Radio	£3.50
BP258	Learning to Program in C	£4.95
BP259	A Concise Introduction to UNIX	£2.95
BP260	A Concise Introduction to OS/2	£2.95
BP261	A Concise Introduction to Lotus 1-2-3 (Revised Edition)	£3.95
BP262	A Concise Introduction to Wordperfect (Revised Edition)	£3.95
BP263	A Concise Introduction to dBASE	£3.95
BP264	A Concise Advanced User's Guide to MS-DOS (Revised Edition)	£3.95
BP265	More Advanced Uses of the Multimeter	£2.95
BP266	Electronic Modules and Systems for Beginners	£3.95
BP267	How to Use Oscilloscopes & Other Test Equipment	£3.50
BP269	An Introduction to Desktop Publishing	£5.95
BP270	A Concise Introduction to Symphony	£3.95
BP271	How to Expand, Modernise & Repair PC's & Compatibles	£4.95
BP272	Interfacing PC's and Compatibles	£3.95
BP273	Practical Electronic Sensors	£4.95
BP274	A Concise Introduction to SuperCalc5	£3.95
BP275	Simple Short Wave Receiver Construction	£3.95
BP276	Short Wave Superhet Receiver Construction	£2.95
BP277	High Power Audio Amplifier Construction	£3.95
BP278	Experimental Antenna Topics	£3.50
BP279	A Concise Introduction to Excel	£3.95
BP280	Getting the Most From Your PC's Hard Disk	£3.95
BP281	An Introduction to VHF/UHF for Radio Amateurs	£3.50
BP282	Understanding PC specifications	£3.95
BP283	A Concise Introduction to SmartWare II	£4.95
BP284	Programming in QuickBASIC	£4.95
BP285	A Beginners Guide to Modern Electronic Components	£3.95
BP286	A Reference Guide to Basic Electronics Terms	£5.95
BP287	A Reference Guide to Practical Electronics Terms	£5.95
BP288	A Concise Introduction to Windows3.0	£3.95
BP290	An Introduction to Amateur Communications Satellite	£3.95
BP291	A Concise Introduction to Ventura	£3.95
BP292	Public Address Loudspeaker Systems	£3.95
BP293	An Introduction to Radio Wave Propagation	£3.95
BP294	A Concise Introduction to Microsoft Works	£4.95
BP295	A Concise Introduction to Word for Windows	£4.95
BP297	Loudspeakers for Musicians	£3.95
BP298	A Concise Introduction to the Mac System & Finder	£3.95
BP299	Practical Electronic Filters	£4.95
BP300	Setting Up An Amateur Radio Station	£3.95
BP301	Antennas for VHF and UHF	£3.95
BP302	A Concise Users Guide to Lotus 1-2-3 Release 3.1	£3.95
BP303	Understanding PC Software	£4.95
BP304	Projects for Radio Amateurs and SWLs	£3.95
BP305	Learning CAD with AutoSketch	£4.95
BP307	A Concise Introduction to QuarkXPress	£4.95
BP308	A Concise Introduction to Word4 on the Macintosh	£4.95
BP309	Preamplifier and Filter Circuits	£3.95
BP310	Acoustic Feedback – How to Avoid It	£3.95
BP311	An Introduction to Scanners and Scanning	£4.95
BP312	An Introduction to Microwaves	£3.95
BP313	A Concise Introduction to Sage	£3.95
BP314	A Concise Introduction to Quattro Pro	£4.95
BP315	An Introduction to the Electromagnetic Wave	£4.95
BP316	Practical Electronic Design Data	£4.95
BP317	Practical Electronic Timing	£4.95
BP318	A Concise User's Guide to MS-DOS 5	£4.95
BP319	Making MS-DOS Work for You	£4.95
BP320	Electronic Projects for Your PC	£3.95
BP321	Circuit Source – Book 1	£4.95
BP322	Circuit Source – Book 2	£4.95
BP323	How to Choose a Small Business Computer System	£4.95
BP324	The Art of Soldering	£3.95
BP325	A Concise Users Guide to Windows3.1	£4.95
BP326	The Electronics of Satellite Communications	£4.95
BP327	MS-DOS One Step at a Time	£4.95
BP328	More Advance Uses of Sage	£4.95